Advancing Global Bioethics

Volume 8

The book series Global Bioethics provides a forum for normative analysis of a vast range of important new issues in bioethics from a truly global perspective and with a cross-cultural approach. The issues covered by the series include among other things sponsorship of research and education, scientific misconduct and research integrity, exploitation of research participants in resource-poor settings, brain drain and migration of healthcare workers, organ trafficking and transplant tourism, indigenous medicine, biodiversity, commodification of human tissue, benefit sharing, bio-industry and food, malnutrition and hunger, human rights, and climate change.

More information about this series at http://www.springer.com/series/10420

Péter Kakuk
Editor

Bioethics and Biopolitics

Theories, Applications and Connections

Springer

Editor
Péter Kakuk
University of Debrecen
Debrecen, Hungary

ISSN 2212-652X ISSN 2212-6538 (electronic)
Advancing Global Bioethics
ISBN 978-3-319-66247-3 ISBN 978-3-319-66249-7 (eBook)
DOI 10.1007/978-3-319-66249-7

Library of Congress Control Number: 2017953260

Printed on acid-free paper

This Springer imprint is published by Springer Nature
The registered company is Springer International Publishing AG
The registered company address is: Gewerbestrasse 11, 6330 Cham, Switzerland

Contents

Introduction

Although biopolitics has a long and diverse history going back to the first decades of the twentieth century, the contemporary usage seems to be dominated by two types of meanings. The term "biopolitics" is either used as a philosophical or a sociological concept relying on the works of Foucault, and his followers, or interlocutors like Agamben, Hardt, Negri, Esposito, Rabinow, and Rose, who mainly focused on the contemporary style of governing populations through biopower, or the term is used more like an umbrella concept referring to public policies regarding the applications of biotechnology and the life sciences. With both meanings at play, biopolitics became a prominent and widely used category both in the philosophical domain and in the ethical-legal discussions of contemporary issues in medicine, life sciences, and society. During the same period, when biopolitics started to develop its intellectual carrier, bioethics was born as a new academic field, and it also became increasingly interdisciplinary and since then ever more politicized. Bioethical issues figure in presidential campaigns and parliamentary elections.

Bioethicists are advisors for governments and frame recommendations for public policies. At least seemingly, the topics, issues, or perspectives of bioethics and biopolitics are overlapping for a large extent, as both fields' representative authors are discussing the promotion of medical research, policies on the AIDS epidemic, regulation of abortion, reproductive rights and freedom, advance directives, commodification of the human body and its parts, or the effects of biotechnological inventions on society. Thus, it seems that bioethics and biopolitics have become parallel but also deeply interwoven fields and activities.

Bioethics might be more easily characterized as an academic field that has a place in universities, especially in the teaching programs of medical schools – with a group of academic experts recognizing bioethics as their professional identity. Although bioethics as a professional identity is not a wide global phenomena, its prevalence and acceptance is more robust in the USA and in the UK, generally in the Anglo-Saxon world, but it is more-more recognized as a profession in European and Asian countries, as well. Within the institutional dimensions of the academia, biopolitics is lacking this professional identity; researchers working on those issues would not define themselves as being "biopoliticians." Biopolitics (re)emerged as a critical concept from the work of Foucault, as a highly genuine theoretical perspective that can be easily differentiated from any common understanding of bio and politics. Both bioethics and biopolitics were born and initially developed around the same decade, in the 1970s, but with significant differences in the career paths that followed. Soon after bioethics as a term was coined, the field easily found its way to medical education and also to wide public debates and political decision-making. Biopolitics had different routes, located more in the theoretical activities of academic researchers, with some distance from everyday activities of wider public.

Those early decades in the development of bioethics were dominated by a rather narrow view on what specifically the *ethics* means after the *bio* prefix. The discipline built up its social career as taking the place of medical ethics that – as a renewed or modernized version of professional ethics – moves beyond ethical codes and gives some place to moral philosophy. However, this early version of bioethics became increasingly problematic because of its theoretical and moral orientation, its methods, and its normative framework which has been challenged as early as the 1990s labelling it a minimalist ethics. According to some contemporary criticisms, mainstream bioethics could be described as a result of a neoliberal ideology of contemporary biopolitics. It is dominated by an autonomy stressing philosophy that is a result of a specific historical and cultural tradition that prioritizes individual autonomy: decisions regarding health and disease should be freely chosen and harm to others should be avoided. Social environment, power relations, and common good became secondary aspects of a moral perspective that celebrates free moral choice. One of the challenges of critical or global bioethics is how to renew the bioethical agenda for incorporating these silenced aspects of moral deliberations. One might argue that we already made some significant steps toward that direction, as we already have an established platform for this broader and more critical bioethics that can be termed as global bioethics. Vulnerability, social responsibility, and benefit sharing are all becoming more prevalent notions in bioethical discussions, which is certainly a critical change in the global transformation of mainstream bioethics.

This book intends to provide a place for interdisciplinary scholarship on these fields based on the firm conviction that problems of bioethics and biopolitics, situated on the intersection of biomedicine, technoscience, and society, can only be properly understood through a diversity of overlapping or even competing theoretical perspectives. So far there were not much academic discussions on the relationship of bioethics and biopolitics, and as one of those initial steps toward such an approach, this book would like to provide a more general orientation than a systematic overview on those issues. These were the ideas that created the background for collecting the chapters of this book, which hopefully provide the reader with a fruitful perspective with a multidimensional philosophical assessment on the relationship between bioethics and biopolitics.

Authors of this book have a variety of background; they are philosophers, doctors, legal scholars, historians, or bioethicists whose approaches can be categorized into different domains that are representative of the contemporary usage of biopolitics as a concept. On the one hand, some authors develop a clear and up-to-date understanding of the primary sources of the concept and related theories of

Agamben, Negri, or Foucault and their potential relevance for bioethics. Another group of authors are applying the philosophical concepts and theories of biopolitics (biopower, *homo sacer*, biocitizenship) on very specific currently debated bioethical issues or using these conceptual tools in order to identify new issues and research perspectives. Some scholars rely on the more mundane understanding of (bio)politics and investigate how its relationship with bioethics could be philosophically conceptualized or follow a more legally oriented analysis on the effects of contemporary biopolitics on human rights and law.

Emerging Concepts of Biopolitics

Compared to the prevalence of using biopolitics as an interpretative tool in important contemporary public debates, the variations in its usages and the diversity of competing theoretical perspectives it offers were not analyzed, described, and assessed with similar intensity and frequency. Among the few exceptions, one should certainly mention Lemke's work (*Biopolitics: An Advanced Introduction*, 2011) that attempts to provide us with a systematic overview of the history of the notion of biopolitics, and this book also approaches to explore its contemporary relevance for some of our important debates. Lemke's book briefly and importantly identifies the often overlooked historical dimensions of the notion of biopolitics and clearly differentiates the philosophy behind those conceptually different approaches. At first sight the word itself has a pretty simple outlook with using the bio prefix to politics, but this simplicity conceals a diversity of meanings and important historical variation in its usages. Lemke first builds up a contrast between the *naturalistic conceptions* and what he calls *politicist concepts*. The former usages emphasize the biological underneath our political activities providing a basis on which politics can be practiced. The latter one is focusing on life processes as the object of politics. This distinction enables Lemke to group together some heterogeneous ideas, most of which were present in the first half of the twentieth century. The list goes from the organicist views of the state around the 1920s to the national socialist state ideology of racial hygiene in the 1940s and ends up at some social Darwinist and biologistic ideas present in contemporary political science. The other notion – biopolitics as the subdiscipline of political practice, the so-called politicist usages – with its main focus on the regulation and steering of life processes emerged around the 1960s. At the beginning, the politicist view of biopolitics appears with the ecological concepts that were focusing especially on the preservation of the natural environment, and then it became refashioned with a technical reading of biopolitics that advocates production and development. Lemke sees both naturalist and politicist concepts of biopolitics as a theoretical failure, as they both see "life" and "politics" as isolated phenomena and both are unable to account for their relationality and historicity.

> The impact of biotechnological innovations has demonstrated that life processes are transformable and controllable to an increasing degree, which renders obsolete any idea of an intact nature untouched by human action. Thus, nature can only be regarded as part of nature-society associations. At the same time, it has become clear that biopolitics also marks a significant transformation of politics. Life is not only the object of politics and external to political decision-making; it affects the core of politics – the political subject." (Lemke 2011, p. 4)

This is the reason why most scholars regard Foucault as the first author able to develop a real theoretical innovation in his concept of biopolitics – where life denotes neither the basis nor the object of politics – that lets us concern life and politics as interrelated and historically developed phenomena. According to Foucault, the birth of biopolitics signals a break in our political order, a fundamental shift both in the practice of politics and living.

Foucault's Legacy

Because of the above-described important role Foucault's thought plays in understanding biopolitics in all its contemporary diversity, this volume departs with an "introductory" chapter that focuses on Foucault's legacy. The first chapter written by Ádám Takács begins by reminding us to consider the very significant functions and sometimes peculiar life of scientific and philosophical concepts that are described being more than ordinary terms or words in our dictionary that simply designate something and have a predetermined meaning, but beyond these functions philosophical concepts acquire a unique innovative dimension. Concepts are also "capable of revealing hitherto unseen scientific problems, motivating the emergence of new ideas, and calling attention to as yet undetected empirical registers" (Takács, Chapter 2). There seems to be a consensus that this innovative conceptual dimension of biopolitics was fully developed by Michel Foucault in his works written in the 1970s. As Lemke pointed out, in the case of Foucault, adding the prefix "bio" to politics is not an innocent extension, but signifies a rupture with traditional political thought: "biopolitics does not supplement traditional political competencies and structures through new domains and questions. It does not produce an extension of politics but rather transforms its core, in that it reformulates concepts of political sovereignty and subjugates them to new forms of political knowledge" (Lemke 2011, p. 33). Biopolitics signifies the constellation in which human and natural sciences develop and normative concepts emerge, structuring political action and determining its proper goals. On the other hand, biopolitics was not a neologism created by Foucault, but as Esposito, and also Lemke, highlights, it was an already existing term since the 1920s that acquired a variety of usages in the subsequent decades of the twentieth century (Esposito 2008; Lemke 2011). Moreover, to add some further layers to this diversity, the chapter refers to another work, in which the competing meanings of the concept from the 1970s onward are described as the term "biopolitics" was serving different theories in political science (Liesen and

Walsh 2012). Following Esposito's diagnosis on the diversity of contemporary usages, Takács reminds us that the more recent carrier of the concept of biopolitics certainly has advantages but risks as well and asks: "Should one abandon this concept in favor of a more accurate vocabulary? Or should one try to delimit its fields of application and insist on an exclusive definition capable of restricting other usages?"

As the title of this introductory chapter suggests, it aims to reconstruct the Foucauldian concept of biopolitics and biopower as a necessary first step in developing a current understanding and orientation in this conceptual diversity. After the publication of his early historical investigations that are more easily situated within traditions of writing the history of ideas, Foucault's interest turned to the question of power. With the publication of *Discipline and Punish*, a new dimension appeared in Foucault's analytic focus – power and power relations – that enabled the development of his concept of biopolitics. According to Foucault, power should not be identified simply with sovereignty, with someone or groups who rule others, but power is represented as a feature of all relations; power is everywhere. It is not centrally located as a point in the middle of a circle, but more like distribution points within a network. Moreover, power does not appear just in forms of negation and negativity, prohibition, constraint, coercion, violence, and physical force but also appears as forms of creation and creativity as something that makes things possible, enables a particular way of thinking, and drives action. The concept of biopower (biopouvoir) is closely related to biopolitics. The historical emergence of biopower is described as three axes of problematizations, where biopolitics appears as a strategy (on individual's body and later on collective entities, like race or populations), as a social rationality (governmentality), and also as political practice (techniques of social classification, calculation, organization via "state sciences"). After introducing the reader with the development of Foucault's concept of biopolitics, the paper turns to the question how far we might use these Foucauldian insights as tools to understand current biopolitics. The second half of the chapter investigates the abovementioned three theoretical axes in order to detect transformations and transfigurations within the contemporary biopolitical condition.

Applications: Contemporary Biopolitics and Bioethical Issues

The second part of the volume presents the reader with a diversity of philosophical perspectives applying the conceptual tools of biopolitical discourses. The common feature of these approaches could be described as their application of the conceptual resources of the theoreticians of Foucauldian biopolitics, with reliance on authors like Agamben, Esposito, or Rose.

The first chapter of this part, written by Ignaas Devisch, opens up with reference to a highly contrasting and historically distant advertisement, a dietary supplement that makes you look fat that was called "fatten U food." This remarkable dietary supplement that was backed up with the slogan "don't look like the poor unfortunate" is the product of the end of the nineteenth century. If we simply check a magazine now, we can find an endless list of wonder diets, but this time the task is not to get plump, but to lose weight. It is certain that in the described period health discourse has changed dramatically. But what exactly happened between these periods? The chapter applies Foucault's concepts and his theoretical perspective to shed light on the fundamental changes of health discourse. With reference to Foucauldian concepts like "governmentality," "pastoral power," or "population," the chapter focuses on the question of how the care for public health became indeed a central task for politics and for all of us. Health is not only on the political agenda of many governments, its presence runs as a (bio)imperative through the whole of society. It appeals to all of us (a totalizing technique) and to each one of us (individualized). There is obviously something wrong with us (pastoral power) if we don't obey this imperative. The second chapter applies a Foucauldian analysis to the contemporary discourse on health promotion, the growing interest of the government and insurance companies in our daily activities, and their attempt to interfere in it. Analyzing the case of obesity, Devisch explores how the discourse on patient empowerment is actually the ethico-medical way through which people are governed. Based on his analyses, in the concluding paragraphs, Devisch summarizes his insights and attempts to searching for answers to his question: How do we choose what we choose? As subjects of empowerment and individuals under control, how far are we responsible for our choices?

Vilhjálmur Árnason – the author of the volume's second chapter – in some sense continues to raise questions of human freedom and determination in biopolitical theorics with building on a contrast between two notions of citizenship using the context of deCODE genetics 2014 biosample initiative in Iceland. He discusses Rose's and Novas' notion of biological citizenship which exemplifies a rich sociological analysis of how developments in biotechnology are shaping contemporary citizens. Árnason highlights that Rose's and Novas' approach is largely framed in the discourse of production and marketing strategies of economy and biology where the vitality of citizens is harnessed for the promise of creating health and wealth. The chapter critically evaluates that notion of biological citizenship, arguing that it ignores the important normative dimension embedded in our traditional notions of citizenship and scientific literacy. However, moving beyond these criticisms, Árnason argues that the opposing approaches supported by these two notions are not to be reconciled because their value consists in the tension between them and the mutual critical resistance that they provide to one-sided analyses of complex phenomena. The chapter uses the example of the Icelandic health sector database project to demonstrate how both these visions of citizenship can be relevant in an actual public policy.

The third chapter uses Agamben's central concept of *homo sacer* in order to provide a philosophical interpretation on the CDoC patient's relatives', caretakers', or nurses' difficulty in creating a meaningful social relationship with their patient who – as being diagnosed with a type of chronic disorder of consciousness – is not, or only intermittently aware, lacking normal awareness. Beyond the specific diagnostic challenges of the variety of conditions of patients with CDoC, the real difficulty lies in clearly articulating the nature of the patient's condition meaningfully, situating it within our everyday experiences, providing a place to CDoC in our moral world. They are neither dead nor alive, neither conscious nor unconscious, and neither an agent nor a passive object, and the link between their pre-trauma and post-trauma identity might be in doubt. The chapter proceeds by first giving us an overview on the variety of conditions that can be grouped together as CDoC, and then it clarifies Agamben's key terms of *zoe*, *bios*, and *bare life* and explicates the philosophical idea of *homo sacer*. The chapter explicitly uses and further explicates the examples of vignettes used in Agamben's work, like the Karen Quinlan case, *coma depasse*, or the figure of Muselmann. Andrew Edgar argues that CDoC poses us the following biopolitical challenge: how to facilitate a creative and poetic role of those who bear witness to the patient's experience and how to articulate for such patient's a radically different ethical status.

The fourth chapter, written by Gardar Arnason, looks on the emergence of left-handers as a social problem from a Foucauldian perspective. The paper provides the reader with an interesting overview regarding the prescientific or premodern notions of left-handedness. The need for the scientific study of left-handedness started somewhere in the twentieth century, which is one of the first signs that it has become regarded as a problem to be scientifically studied. Although significant energies were poured into attempts to acquire an adequate scientific understanding of left-handedness, still today, the exact causes of left-handedness or more broadly the phenomena of human lateralization remain poorly understood. He relies on specific dimension of biopolitics defined as management of population with asking the question of how the government of a certain group of people requires or uses scientific knowledge. As such in a "biosocial group," the case of left-handers is used as an example. Parents, teachers, school principals, and a variety of experts had to deal with "the problem of left-handedness," and they called upon scientists to give them more knowledge about these problematic children. Especially psychology and health sciences provided a vast amount of research on this particular group in the past century from a variety of perspectives. The chapter discusses one such perspective, and a series of studies aimed at answering the question whether left-handers die on average younger than right-handers. Discussing these studies, Arnason attempts to demonstrate how problematic the scientific definition of left-handedness is and also what power effects this sort of research has on its subjects.

The fifth and last chapter of "applications" part of the volume is written from the perspective of human rights and biolaw. Judit Sándor follows the concept of molecularization in discussing how biotechnological developments are posing unseen challenges to legal norms. Molecularization was coined by Nikolas Rose with reference to the newly emerged phenomenon of the last decades that approached the fact that beyond the individual human body being one of the primary sites of biopolitics, more and more biopolitics is being extended to parts or fragments of the human body. Sándor discusses the practical implications of molecularization and explores how the new biopolitical thinking determines legal structures and public discussion in light of the current advances in biotechnology. As a result of the biotechnological breakthroughs of the last decades, followed by their rapid translation to every applications, currently human DNA, mitochondria, cells, cell lines, and gametes are defrosted, tested, harvested, diagnosed, and edited in laboratories in an increasing scale. The multiple form of interventions on human organs, body parts, cells, and molecules resulted not just in vibrant public debates about the ethical and political issues but became rather challenging targets of legal policymaking. The focus of this chapter is to discuss human rights as the catalyst and synthetizing force in these debates. Human rights, providing rights to the persons, are now extended with open boundaries, such as rights related to human tissues, to human DNA, to brain-dead persons, and to in vitro embryos. As a result of these developments, law reinterprets the notion of "human" on the molecular level. The chapter explicates a list of famous court cases that had a determining role on how we might elaborate the legal status of these new molecular entities, like the Moore, Hagiwara, Diamond v. Chakrabarty, and Pratten v. British Columbia cases.

Connections: Bioethics and Biopolitics

The third part of the volume builds on three papers that explicitly reorient the focus of discussions from the various applications of biopolitical theories on bioethical issues toward directly questioning the relationship between bioethics and biopolitics. The first chapter is written by Henk Ten Have under the following question: Can bioethics escape from biopolitics? In the analytic perspective of this chapter, bioethics and biopolitics are represented as two discourses that might have developed as related fields, but according to Ten Have's diagnoses, bioethics became the dominant manifestation of biopolitics itself: "Bioethics has generated a political regime, a system of governance that administers, fosters and secures life as a form of 'biopower', controlling the population and disciplining the individual." Ten Have's description of mainstream bioethics makes him enable to establish a close link with the ideology of neoliberalism as it emphasizes internal regulation by autonomous subjects rather than external force and pressure. The prime figure of bioethics is the rational, autonomous individual who is the entrepreneur of himself (Lemm and Vatter 2004). Bioethics as the manifestation of biopolitics emphasized the value of humanitarianism, respect for individual autonomy, as the center for ethics and politics. However, at the same time it could not but neglect the social, political, and economic dimensions of human life. Ten Have argues for reestablishing the initially present critical potential in bioethical discourse, instead of simply reiterating and facilitating biopower with the uncritical use of an autonomy stressing philosophy of neoliberalism. This critical tradition of bioethics can be brought back by bioethics' rebirth as global bioethics. The envisioned roadmap of the chapter depends on our answer to the question whether bioethics can "develop a moral discourse, using the insights of biopolitical reflection (particularly the views of Didier Fassin, Roberto Esposito and Melinda Cooper) that criticizes, resists and opposes biopower."

Tuija Takala authored the second chapter in part three of the volume, which discusses the politics in bioethics in its traditional, more mundane sense, with a focus on how bioethics' relation to political activities can be described. According to

Takala, bioethics has never been far removed from politics or political agendas. Besides the purely theoretical, conceptual, or descriptive approaches, everything in bioethics is political, as it is about rights and responsibilities, justice, and entitlement – all of which are political notions. The chapter distinguishes between speculative and practically relevant, purely theoretical and policy-oriented types of bioethical inquiries. The former represents the hope of pure academic bioethics as the latter always shows the dirty world of politics. Although the world might need practical solutions, "compromising on the theoretical, methodological and conceptual levels means that bioethics loses its scientific justification and comes closer to bioethics." Takala uses some dominant approaches in bioethical inquiry concerning the ethical issues of new technology examples in order to highlight primary forms of bioethical reasoning, forms of analyses, and argumentation (slippery slope argument, "playing God," naturalness, "bad Samaritan argument"). At the end of the day, some might have difficulty in distinguishing bioethics as a political activity from bioethics as an academic endeavor. In much of what we see in contemporary bioethics, scholars simply "market their own moral and political beliefs by seemingly clever arguments." On the other hand, bioethics' strategies of demarcation are constantly challenged by the increasing presence of interdisciplinary approaches in bioethical research. Although applying an interdisciplinary approach on practical moral issues allows us to bring together a more substantial body of knowledge for understanding the related questions, the theoretical and methodological foundations on which our final answers will be based will be highly questionable. Thus, Takala argues that assessing the scientific validity of interdisciplinary contributions to bioethics is difficult, and also, as those are accepted as contributions to bioethics literature, "they tend to expand the fields rather than advancing it."

The final chapter, written by Darryl Gunson, discusses the conceptual and practical relationship between bioethics and biopolitics. The chapter departs with describing Lemke's account on bioethics that serves as an object of critical refusal at the end of the paper's argumentation. The chapter demarcates certain ideal types regarding the nature of bioethics which could illustrate two ends of the spectrum upon which discussions of the field of bioethics occupy according to the following criteria: "the first concerns the scope of bioethics, which includes the aims of bioethics and its problems; the second involves the details of the method(s) of bioethics including the use of empirical data." Regarding the scope of bioethics, the paper discusses two suggested models which offer a broader and narrower focus for bioethics. According to Gunson, the first model suggests that bioethics is, or should be, a global undertaking, with a broader and renewed focus on global justice, basic health needs, and our relation to the environment. And the second model suggests that bioethics' primary focus should be on problems with healthcare systems and issues emerging from the advancement of biotechnology. The paper continues by comparing these models with some dominant characterization of biopolitics. Gunson argues that the relationship between bioethics and biopolitics is that they focus on nearly identical social issues, but biopolitics is broader and deeper, focusing on the wider political context of policy and regulation, and that this is not part of the remit of bioethics. According to Gunson, this judgment has some plausibility only if we follow a narrow definition of bioethics. "Once we shift our focus in bioethics to appreciate the necessity for broadening our horizons – perhaps to a 'global bioethics' – and once we understand the necessity for empirical research from all relevant sciences, we begin to see that bioethics does have a legitimate concern with the type of questions and analyses hitherto associated with the field of biopolitics." The chapter describes how the two biofields of academia are related and argues for a more critical bioethics that uses the insights of social sciences, historical research, and biopolitical theories.

The idea of publishing this volume emerged from discussions during the organization of the 28th conference of the European Society for Philosophy of Medicine and Healthcare, which took place in August 2014, at the University of Debrecen, Hungary. Our basic aim with organizing the society's yearly meeting under the title of "Bioethics and Biopolitics" was to invite our colleagues to reflect on various aspects of these two important concepts, fields of knowledge and academic activities, with focusing on their interrelatedness. I hope the volume will find its audience and the readers find the volume both exciting and useful in providing a detailed understanding regarding the various contemporary activities running under the title of bioethics and biopolitics.

I cordially thank the editors of this Springer series for their invitation to develop this book project, and I'm especially thankful to my great colleagues, the authors of these chapters.

Debrecen, Hungary Péter Kakuk

Bibliography

Agamben, Giorgo. 1998. *Homo sacer: Sovereign power and bare life*. Trans. D. Heller-Roazen. Stanford: Stanford University Press

Beauchamp, Tom L., and James F. Childress. 2013. *Principles of biomedical ethics*, 7th ed. New York: Oxford University Press.

Esposito, Roberto. 2008. *Bíos: Biopolitics and philosophy*. Trans. T Campbell. Minneapolis/London: Minnesota University Press.

Fassin, Didier. 2009. Another politics of life is possible. *Theory, Culture & Society*; 26(5): 44–60.

Foucault, Michel. 2008. *The birth of biopolitics: Lectures at the Collège de France 1978–1979*. Trans. G. Bucknell. London: Palgrave Macmillan.

Hardt, Michael, and Antonio Negri. 2000. *Empire*. Cambridge, MA: Harvard University Press.

Lemke, Thomas. 2011. *Biopolitics. An advanced introduction*. New York/London: New York University Press.

Lemm, Vanessa, and Miguel Vatter (eds.). 2004. *The government of life. Foucault, biopolitics, and neoliberalism*. Bronx: Fordham University Press.

Liesen, Laurette & Mary Barbara Walsh. 2012. The competing meanings of "biopolitics" in political science: Biological and postmodern approaches to politics. *Politics and the Life Sciences*, 31(1–2): 2–15.

Rabinow, Paul, and Nikolas Rose. 2006. Biopower Today. *BioSocieties* (2006) 1, 195–217.

Rose, Nikolas. 2007. *The politics of life itself: Biomedicine, power, and subjectivity in the twenty-first century*. Princeton: Princeton University Press.

Ten Have, Henk, and Bert Gordijn. 2013. *Handbook of global bioethics*, 4 volumes. Dordrecht: Springer.

Part I
Concepts: Biopolitics and Biopower

Chapter 1
Biopolitics and Biopower: The Foucauldian Approach and Its Contemporary Relevance

Ádám Takács

Abstract The notions of "biopolitics" and "biopower" enjoy commonsensical plausibility in many fields of humanities today. From philosophy and sociology through cultural and gender studies up to various forms of contemporary political thinking, these notions are used and reused in many descriptive and normative approaches. However, even if it is often highlighted that the work of the French historian and social theorist, Michel Foucault served as a cornerstone in attaching the prefix 'bio' to the words 'politics' and 'power', the question as to for what reasons these terms, designed originally for historical research in Foucault, could reach such an interdisciplinary popularity remains to be worth studying. With this context in mind, this paper has two objectives. On the one hand, it seeks to reconstruct the meanings and roles of the notions of biopolitics and biopower as they are displayed in Foucault's historical and theoretical researches. On the other hand, it aims to foreground the theoretical significance as well as the descriptive and normative values that could be associated today to these notions in various fields of humanities within the contemporary conjuncture of biopolitical thinking.

Keywords Foucault • Biopolitics • Biopower • Problematization • Historical research

1.1 On the Fortunes of the Concept of Biopolitics

Concepts in social sciences and humanities, just like social scientists and intellectuals themselves, are capable of making spectacular careers. Concepts can develop over time, they can be enhanced and improved, often outside the field where they were born, and become even fashionable. This also means that the specific point in time and particular field of research that marked the beginning of a concept's usage are far from being the only decisive factors to determine its further development and

Á. Takács (✉)
Eötvös Loránd University, Budapest, Hungary
e-mail: takacsadam@hotmail.com

P. Kakuk (ed.), *Bioethics and Biopolitics*, Advancing Global Bioethics 8,
DOI 10.1007/978-3-319-66249-7_1

application. In fact, scientific concepts are not only interdisciplinary entities per se, but are also open to usages and procedures that may pay little or no attention to the original intentions that initially brought the concept into play. For a scientific concept is not simply a word or symbol equipped with a predetermined meaning. Even if it requires definition, such concept operates in the scientific discourse as a form of concentrated information, the value of which is strongly dependent upon its concrete unfolding and context. Consequently, what makes the scientific career of a concept is never simply the succession of consequent applications. Apart from being applied in descriptions or interpretations as instruments or representations, concepts are also capable of revealing hitherto unseen scientific problems, motivating the emergence of new ideas, and calling attention to as yet undetected empirical registers. In addition, conceptual work can also help to redesign social reality itself insofar as scientific concepts tend to be used and reused as representations in the processes of social negotiations, individual or collective self-identification or in political interactions (Bourdieu 1992).

The current fortunes of the term "biopolitics" aptly demonstrate what is at stake when a scientific concept takes on a life of its own via the creative process of its applications. Indeed, the current appeal of this concept in a number of fields has led to not only an extensive use that reaches beyond the sphere of academic scholarship, resurfacing in media and public discourse, but also the emergence of broad and competing interdisciplinary connotations. As of late, philosophy, history, political science, gender, cultural and public policy studies, sociology, and various bio- and life sciences have begun to make systematic use of this notion, adapting it to the specificities of their own theoretical and research purposes. As a consequence, the notion of biopolitics has, by now, become the object of a growing field of discursive appropriation with far-reaching scientific, political, and social implications. For this very reason, the term has been endowed with a complex conceptual profile that is far from being governed by a singular definition and usage.

Thus even if the canonical contexts and uses of the term "biopolitics" are mostly rooted in the field of contemporary moral, political, and social philosophy (Losoncz and Takács 2015), the concept has in the meanwhile acquired such a versatile character that makes theoretical studies carried out under its label to cover a wide range of topics and scientific areas. In philosophically driven approaches, biopolitics has opened up space for new discourses on the human condition that address questions of morality, human agency, and sociality in terms of the bodily constitution and physical vulnerability of man (Agamben 1998). In studies with a focus on historical and social issues, interest in biopolitics typically translated into works on the principles and various forms of managing human population via the discursive and legal regulations of public hygiene, public health, gender, sexuality, or race (Rose 2007). Discourses on biopolitics also contributed to the renewal of the toolbox of political analysis, a process that led to the reconceptualization of power as the constitutive form of control over bodies in modern and contemporary societies (Hardt and Negri 2000). Finally, the topic of biopolitics provided a space in which the discourses of the bio- and life sciences and those of the social and human sciences can mutually influence each other. Within this context, biopolitics has often been considered a mode of analysis that can liberate life sciences and bioethical discourse from their abstract and decision-oriented nature and embed them within a larger historical and epistemological context (Lemke 2011).

Nevertheless, a successful conceptual career can also turn into the source of certain confusion. The question can be asked whether a very broad potential of use of a given concept might not end up weakening its descriptive and explanatory power in general. Indeed, it could be argued that when a concept takes on too many—and at least partly contradictory—meanings, its value as an analytical instrument is likely to decrease. Likewise, in the case of biopolitics, one can wonder whether the robustly interdisciplinary use of this term might not run the risk of generating severe conceptual inflationary effects. Roberto Esposito, one of the leading figures of contemporary Italian philosophy f biopolitics has recently called attention precisely to this problem. In his book *Bíos: Biopolitics and Philosophy*, Esposito not only argued that the term biopolitics "has opened a completely new phase in contemporary thought," and that it made "the entire frame political philosophy emerge as profoundly modified," but also concluded that "Far from having acquired a definitive order, the concept of biopolitics appears to be traversed by an uncertainty, by an uneasiness that impedes every stable connotation". (Esposito 2008: 13–14).

If Esposito is right in his diagnose then the question should be asked as to what extent researchers should take this situation into account. What would be the best strategy to cope with the fact that drawing upon the notion of biopolitics as a descriptive or explanatory category is exposed to the risk of generating loose meanings? Should one abandon this concept in favor of a more accurate vocabulary? Or should one try to delimit its fields of application and insist on an exclusive definition capable of restricting other usages? Or again, should one simply try to fabricate a new working definition of this notion that stands on its own and is better suited for a specific purpose?

It is with this context in mind that the present study seeks to revisit the work of the French historian and philosopher Michel Foucault. If Foucault's work can be considered important and to some extent exemplary for today's debates over biopolitics, this is not only because it were his writings that essentially introduced this term into the vocabulary of social and human sciences some 40 years ago.[1] Foucault's significance also lies in the impact his historical and theoretical approach had on current theories of biopolitics in raising awareness of the role of historical information in the conceptualization of present-day issues. The aim of this study is therefore to go beyond seeing Foucault's work as being merely a source of authority when dealing with the question of the current potentials of biopolitical reflections. Rather, the goal of this writing is heuristic and consists in demonstrating how historically informed research, within which the notion of biopolitics can be descriptively elaborated, can be put into the service of a conceptual approach designed to retrace or

[1] It must be noted that the term "biopolitics" was invented in the 1920s (Esposito 2008) and was employed discretely from the 1970s onwards in very different traditions of political and social theory (Liesen and Walsh 2012).

reinvent the biopolitical phenomena in the current situation. Thus, we will argue that Foucault's approach and its actual or potential contemporary applications can offer more than an exclusive definition of biopolitics. It offers a model to understand how drawing upon the concept of biopolitics, while avoiding the pitfalls of generating loose or restricted meanings, can be put in the service of a multiscopic social analysis carried out in the framework of a genealogically driven problematization of our present.

1.2 Foucault and Biopolitics

The emergence in the mid 1970s of the topic of biopolitics in Foucault's work fits with his broader theoretical and historical analysis of social power and power relations. The first time he elaborated on the question of power was in his historical work devoted to the emergence of the mechanisms of imprisonment in eighteenth century European societies. The results of this research were published under the title *Discipline and Punish: The Birth of Prison* (Foucault 1977). Yet despite of what its title suggests, the scope of this book was not limited to providing a historical analysis of the social and theoretical mechanisms behind the massive changes that occurred in the Western penal system during the modern age. Foucault was also concerned with demonstrating the importance of the cultural and rational shifts that led to the dominance in modern societies of the prison as an institution. In doing so, he pointed out the changing nature of the relationship between the various forms of power and the human body. Within this framework, the prison was presented as the institutional manifestation of the emergence of a new and paradigmatic technology of power. This latter was invented as a way of disciplining the mental and physical behavior of social agents, and do so according to a hitherto unseen form of social rationality and calculation that revolved around the control over bodily presence. Yet Foucault also demonstrated that patterns of the same rationality were operative in places such as schools, hospitals, factories and military institutions. Based on such findings, Foucault could conclusively speak of the emergence in modern societies of a "disciplinary" form of power.

It is important to note, however, that Foucault's analysis of power did not revolve primarily around the questions of political rule and domination enacted via instruments of coercion or techniques of legislation. For him, power was not simply an attribute of those who are in charge in a society. Rather, in his various analyses on the topic, Foucault focused on issues through which the phenomena of power were historically manifested in terms of normalization, rationalization, institutionalization, control, subjectivation and embodiment connected to the social life of concrete individuals and communities. Consequently, one is hard pressed to find in Foucault's works the elaboration of a general theory of political authority and rule. Instead, he analyzed power as a way of rationalizing, controlling, or disciplining human behavior and body, and as something that was immanent in the social and political management of relations of production, relations of social organization and administration, family relations, sexual relations, and, in fact, all sorts of relations comprising the social context in general. The key point about Foucault's historical approach to the issue of power is that it essentially transcended the traditional sphere of politics as such and conceived of power as a socialized, institutionalized, and embodied phenomenon.

Within this theoretical framework of Foucault, the notion of "biopolitics" is closely related to the concept of "biopower" (*biopouvoir*), which refers to a specific mode of power relations. This type of power is not without a certain distinct historical and theoretical complexity. In his first attempt to define biopower, Foucault described it as the result of the historical emergence in modern European society of a new form of control over human needs and potentials. Biopower, according to him, is "the set of mechanisms through which the basic biological features of the human species became the object of a political strategy, of a general strategy of power" which process reflects "how, starting from the eighteenth century, modern western societies took on board the fundamental biological fact that human beings are a species" (Foucault 2007: 16). In other contexts, Foucault also spoke about biopower as the merging in modern societies of political and biological knowledge, which constituted a precondition for capitalist development. Due to the emergence of this new constellation of knowledge in eighteenth century Europe, argues Foucault, "Western man was gradually learning what it meant to be a living species in a living world, to have a body, conditions of existence, probabilities of life, an individual and collective welfare, forces that could be modified, and a space in which they could be distributed in an optimal manner. For the first time in history, no doubt, biological existence was reflected in political existence" (Foucault 1981: 142).

Nevertheless, this distinct period in eighteenth century Europe when, in Foucault's view, society reached—through the implementation of new forms of power and knowledge—the "threshold of biological modernity" (Foucault 1981: 143. translation is modified) was not simply the outcome of a spontaneous historical transformation. For Foucault, biopower was not something that happened to society, but rather something that was produced or invented by society itself. In other words, far from being an unstructured social process, the rise of this type of power was the result of the emergence of new social and political strategies that sought to respond to a certain strategic need carved out of social reality itself. Biopower for Foucault therefore names the emergence of a novel type of rationality consisting of evaluations, calculations, conceptualizations, and decisions targeting the biological features of society (Foucault 2003, 2007). But insofar as the knowledge it brought into play responded to concrete ways of problematizing the social, the emergence of this new form of power tended to institute itself on the political level, too. In short, it carved the way for biopolitics as a political practice.

It can be argued that when it came to analyzing the historical emergence of biopower, Foucault identified three axes of problematization, with their respective focus on the emergence of biopolitics as a strategy, as a social rationality, and as a political practice in modern society. The first axis of problematization relates to the invention of a new type of body. Foucault argued that, from the premodern period, it was what he called "the anatomy politics of the human body" (Foucault 1981: 139) that determined the nature of the power over life in most social circumstances. This latter sort of politics was aimed at disciplining, optimizing, and training the individual body in order to increase its efficacy and integrate it into systems of social and economic control. Yet, somewhat later, this form of power was complemented by another form of strategic power, which Foucault called, as we have already seen, the "biopolitics of the population" or the "human race" (Foucault 1981: 139, 2003: 243). Unlike the previous form, biopolitics focused not so much on the individual human as a living being, rather, it sought to recapture the type of body as belonging to the human species as a whole. In this sense, the body became politically interpreted as a collective entity subjected to the biological and social conditions of existence, such as "propagation, birth and mortality, level of health, life expectation and longevity along with all the conditions that can cause these to vary"(Foucault 1981: 139). Paying special attention to these conditions also made it possible to conceive the collective human body as a field of forces that could be modified and optimized. Historically speaking, it was thus the invention of the human population as social body that in Foucault's view marked the fundamental condition for the implementation of the biopolitical strategy in modern societies (Foucault 2007).

The second axis of biopolitical problematization recognized by Foucault relates to the fact that the invention of the population as the possible object of political strategies also resulted in the crystallization of a new way of exercising power. Whereas the mechanisms of power that targeted the individual human body were enacted essentially through disciplinary means (such as subordination, discipline, training, etc.), which Foucault has studied in detail with respect to the institution of the prison (Foucault 1977), biopolitics was introduced and established at the social level in the form of mechanisms of control. Control in this case, however, was far from implying a repressive set of maneuvers. One of the most spectacular results of Foucault's historical analysis concerning the biopolitical invention of the population lies in his demonstration that the population as a social and political entity proved to be essentially unsuited to disciplinary treatments (Foucault 2003). In fact, Foucault noted that the living conditions is subject to processes that either depend on long-term social developments (e.g., level of public hygiene, rate of birth and mortality, conditions of work and living, etc.), or tend to unfold quite outside the realm of human power (e.g., epidemic, diseases, famine, etc.). Consequently, rather than calling for coercive policies, the realization of biopolitical control over the population required the implementation of measures such as classification, regulation, prevention, provision, and maintenance of security. For this reason, it was the liberal form of "governmentality," as opposed to the traditional concept of sovereign power, predominant up to the eighteenth century that constituted in Foucault's view the most prominent form of political rationality giving rise to a biopolitical attitude in the process leading to the formation of the modern state (Foucault 2008).

Finally, the third axis of problematization in Foucault's work is centered on the question of knowledge, mobilized historically by the birth of biopolitics. Since, as he demonstrated, the practical equivalent of inventing the population was to learn how to measure and control it, biopolitics as a social strategy turned out to be entirely dependent on techniques of social classification, calculation, and organization (Foucault 2007). In other words, the question of population could only become a political issue or question of power once it had been sufficiently elaborated as a scientific and technological problem. In this sense, the birth of biopolitics overlapped, argued Foucault, with the emergence of the need for a new type of political and social knowledge about society. Statistics, demography, political economy, and other "state" sciences were called upon to help the state's efforts to measure, control, and treat manage the population. At the same time, medical knowledge from the beginning of the early eighteenth century was extensively mobilized in order to improve the living conditions of the population by advancing public hygiene and combating epidemic diseases and famines. Yet, as Foucault pointed out, the concept of biopolitical intervention as such was largely unthinkable within the medical rationality of the era (Foucault 2007). Medicine back then dealt essentially with the individual and not the collective or social body. It was precisely the biopolitical way of thinking about power and population, along with its social scientific background, that allowed for the transformation of medicine into an effective modality of social intervention. In this manner, the birth of biopolitics also marked the emergence in modern European societies of social medicine, previously called public hygiene, both as a concept and a practice (Foucault 2007).

1.3 Using Foucault in Understanding Current Biopolitics

Using Foucault's historical account of biopower and biopolitics as a conceptual tool or framework for studying contemporary social phenomena makes all the more sense given that he himself designed his analyses with such potential application in mind. Generally speaking, for Foucault, historical reflection has always been the way through which to approach the present for analytical and critical examination (Takács 2004). For him, turning something into a historical object and using this same object as an instrument for analyzing current facts and trends in knowledge and studying the practice of power or other modes of societal behavior was essentially the two sides of the same coin. This is precisely one of the major reasons that keep the "usages of Foucault" constantly on the agenda in a wide range of fields in the humanities and social sciences (Oulc'hen 2014). Nevertheless, such attachment to the research perspectives opened up by Foucault stimulates neither an authoritative closure on topics not discussed by him nor a strict disciplinary commitment. On the contrary, using Foucault's ideas as a toolbox rather than as a doctrinal foundation seems to characterize most analytical approaches that seek to capitalize on his work across scholarly fields and disciplines.

What would a description of our current biopolitical situation from a Foucauldian perspective entail? For obvious reasons, such an exercise cannot simply involve turning Foucault's historical explanation of the emergence of biopolitics into a principal model for understanding our present situation. An attempt like this would not only disregard the changing nature of biopolitics in modern and present-day societies, but would also result in the misapprehension of the possible shifts and displacements in the use of scientific concepts themselves. Thus, whereas a critical reassessment of Foucault's historical account of biopolitics from a post-colonial or gender perspective is certainly possible (Stoler 1995), an innovative approach to today's biopolitical phenomena should go further than revising or expanding his earlier findings. Instead of focusing on the historical data and narrative presented in Foucault's analyses, it is probably more promising to revisit the core problems that informed his analysis of social reality and the emergence of biopolitical practices. Reexamining these axes of problematization from a current perspective would not only allow for an unbiased use of Foucault's ideas and results, but would also make it possible to detect transformations and transfigurations within the biopolitical condition. These three core issues problems, as we have seen, revolve around the questions of the status of the body, the form of biopower, and the nature of knowledge implied in biopolitical practices and strategies, respectively.

1.4 Body

It can be argued that one of the major features of our current biopolitical condition lies in the difficulty of separating sharply from one another the spheres of the individual and the collective apprehension of the human body. Neither considered merely as closed anatomic unit, nor seen as element of a population, the human body tends to acquire in contemporary societies a mixed status. This means that from a cultural or political point of view, the body as a living entity is marked by a certain duality: it effectively functions as a social object when it comes to society, yet at the same time it also fully retains its status as a subjective or personal disposition of the self (Synnott 1993; Rose 1996). The body becomes a social object when it serves as a major location or backdrop for the production of various social meanings (sexual, medical, political, aesthetical, commercial, etc.). Yet it also remains individual or subjective insofar as the production of these meanings induces a specific relation to ourselves as embodied entities. Whereas the body is usually regarded as a bearer of social codes in political and cultural representations (Synnott 1993; Riley 2008), individual embodiment is particularly emphasized in biomedical discourses which consider the human body as a privileged site of natural immunity and vulnerability (Cohen 2009). It is precisely in the crossroad of these individualizing and socializing tendencies where current biopolitical strategies and practices about the value of life and the "politics of life" arise (Rose 2007; Spade 2011).

Furthermore it can be also argued that the emergence of this hybrid biopolitical status of the body was fostered by such social experiences as the human mass destruction carried out by totalitarian regimes during the twentieth century, especially the Holocaust. With this context in mind, and in referring to Foucault's work, Giorgio Agamben argues that the politicization of natural life on the one hand, and the naturalization and animalization of politics, on the other, are the two sides of current productive biopolitical strategies (Agamben 1998). Other motivating factors behind these new biopolitical developments can be related to the new status accorded in our society to sexuality as a political matter. Of particular importance in this respect is the emphasis put in recent queer theory on the steadfast politicization of sexuality, race and affectivity, and the emergence of corresponding new constellations of biopower (Puar 2007; Chen 2012). Also, the global significance of modern sport as one of the most powerful ways of practicing, exposing and evaluating the human body contributes to the creation of blurred zones at the level of bodily experience and the production of meanings. The social, institutional and economic side of sport on the one hand, and its complete reliance on individual embodied capacities, on the other, creates biopolitical situations where standard phenomena related to health issues, physical development, and drug use acquire hitherto unseen social significance.

From a social diagnostic point of view, in all of these cases, the individual or subjective bodily experiences and capacities are inseparable from the shared social values attributed to the body as such. This, however, also shows that by now the target of biopolitical considerations and practices has significantly shifted. The body in our society is neither simply disciplined, nor merely controlled. In addition, it is considered fundamentally as a locus of self-realization and self-concern, the place where social norms are to be interiorized and expressed. Paradoxically, it can be argued that nowadays it is the body, rather than the mind or the spirit in itself, that increasingly functions as the organ of individual and social autonomy (Hardt and Negri 2000, 2009). Nevertheless—and this is where the body becomes the target of newfangled biopolitical strategies and conflicts—autonomy does not cease to encompass two conventional basic social values: that of self-subsistence and that of liberty (Fehér and Heller 1994). It is precisely the conflictual nature of these two values—namely the value of life as continuation and persistence, and the value of life as the locus of the practice of freedom—that essentially undermines the current biopolitical situations pertaining to the living body. In effect, most of the current political debates on abortion and euthanasia, organ trade, use of drugs or doping in sport are captured in this fundamental dichotomy of norms. It is thus the tension between these two basic set of values that marks one of the major axes of problematization that help to understand the ways in which the biopolitics of the body today is brought into play.

1.5 Power

The current biopolitical status of the body is in fact inseparable from the ways by which power operates in several fields of society. The fact that neither disciplinary authority, nor the control of population in the Foucauldian sense seem to adequately describe the current nature of biopower should by no means seen as an indication for the latter's insignificance as a factor in the social scene. On the contrary, integrative tendencies fostering the normalization of the body and life appear as a dominant feature of the contemporary conjuncture of biopower. These tendencies manifest themselves in the spread of normative and regulative mentalities and the corresponding implementation of public policies with regards to issues of health; sex, race and gender; domestic and family affairs; as well as consumption (Turner 1992; Rabinow and Rose 2006). Thus, it can be argued that the emergence of these new forms of biopower is embedded in the functioning of power as a protective force in today's societies (Brown 1995; Esposito 2011). Furthermore, effects of biopower are manifested clearly in the ways by which the question of agency and authority are brought together in the current political, religious and biomedical discourses on the value of life (Agamben 1998; Rose 2007). In this sense, it would not be incorrect to portray this current conjuncture as a sign for the intensification of power and biopower in society (Nealon 2008).

With this context in view, an entirely new element in the working of biopower can be identified in the fact that contemporary culture witnesses the wide propagation of normalization and protective strategies of life in the name of self-worth and self-control. In other words, practices that make the "self" to be the target of regulatory social mechanisms over life constitute today a major element in the social performance working of biopower (Rose 2007). The results of these new strategies can be recognized, for example, in the contemporary normative cult of healthiness and fitness, in the generalized overall tendency of displaying the body as image, and in the spread of the practice of the designed biomedical manipulation of the conditions of the individual body and life. Within this framework, the self is portrayed as the very agency of power which, through its innate and intimate relation to itself, carries out an interpretation of life as a way of living and, if necessary, initiates subsequent modifications in it. Nevertheless, making the self the center of power relations over life does not necessarily weakens its dependency on regulative or exploitive social strategies. Indeed, following the footsteps of Foucault, it could be argued that normalizing technologies ultimately preserve and expand upon the technologies of the self (Heyes 2007). In this sense, the new forms of operative biopower do not simply tend to reinforce the distinction between what is socially valued as normal or abnormal, but—in making the self's capacities reliant upon such evaluations—they also make the repressive or self-expressive elements of biopower virtually indiscernible.

Ultimately, what makes this conjuncture of power challenging from a biopolitical point of view is precisely the fact that the meaning of the self can vary depending on the nature of biopower relations it is associated with. In effect, it could be demonstrated that a number of notions, ideals and practices of biopolitical self subsist simultaneously in our society; so one can speak, respectively, of a subjective, medical, technological or political self, or of those selves targeted by marketing and media. Furthermore, the concurrence of and, at the same time, conflictual relationship between these notions in the social and political field generate exigent moral and social problems. Whether one considers the new trends in the functioning of power over life as a pattern leading to the colonization of the private (Nealon 2008), or as a chance to empower the self against some of the unfair tendencies in public and state policy concerning sexuality and gender (Cooper 1995), accounting for the assessment of the effects of biopower can yield very different theoretical and practical conclusions. In any case, it seems obvious from this perspective that the question of the complex relationship between the embodied self and the management of power constitutes in our age a central axis of biopolitical problematization.

1.6 Knowledge

One of the most striking phenomena in today's biopolitical discourses that unfold in the public, but to some extent also in the political domain, is the unconditional value attributed to medical knowledge. Whereas in the earlier historical periods analyzed by Foucault, biopolitics entailed a predominantly political rationality backed up by various social, medical and public policy concerns (Foucault 2007), medical and biomedical sciences today tend to provide exclusive support for the publicly orchestrated biopolitical strategies and aspirations. What is more, in dealing with the various social issues pertaining to the body and life, medical knowledge is presented as a scientific and institutional authority, the competency of which is very difficult to challenge on its own level and own terms. Within this context, it is not simply the recent growth of medical knowledge in society which is remarkable in itself (Ten Have et al. 1990). More importantly, it is the exclusive medical constitution of the relationship of knowledge and power in the social distribution of health and illness, and its role in establishing the principles of diagnostic and therapeutic normality which constitutes a unique observable phenomenon in the field of biopolitically informed knowledge (Turner 1992).

What makes the monopoly that medical knowledge enjoys today so striking is that it goes against the modern trend that makes socially distributed knowledge to be accessible or challengeable to even those who do not belong to the privileged few with professional competence. It is also remarkable that this exclusive authority over biopolitical knowledge seems to undermine the guiding principles of a different biopolitical tendency, namely that of the perception of the body as an autonomous and self-centered subjective and public entity. This contradiction, for example, appears clearly in recent bioethical discourses in which efforts to harmonize individual, public or professional moral principles and concerns face severe challenges (Taubler 2005). Simultaneously, the privileged position of medical knowledge in today's society also manifestly promotes political strategies that tend to reinvigorate earlier coercive forms of biopolitical interventions. The role of medical knowledge in social policy and the support it provides to the criminalization of certain groups (drug users, for example) reestablishes certain disciplinary aspects of power (Rodger 2008). At the same time the utilization of human genetic information for political and legal purposes leads to the revival of ancient forms of social control (Chriss 2007). In sum, the question of the status of medical knowledge in contemporary society marks an important area where biopolitical considerations as well as conflicts are constantly brought into play.

1.7 Concluding Remarks

Despite its conceptual plasticity and diverse history, the notion of biopolitics is being harnessed today by pioneering empirical and theoretical researches. Biopolitical approaches offer innovative strategies for analyzing the ways in which social and political considerations play an increasingly significant role in shaping what could be described as the biological and bodily features of the human condition. The advantage of this term, along with its underlying principles, lies in the fact that it is able to incorporate various research perspectives and assumptions in a more or less transparent way. The field opened up by biopolitical approaches can equally accommodate empirical, historical, philosophical, and political research methods and theorizations. Within this context, the work of Michel Foucault is a groundbreaking one. In this paper, we argued that the vocabulary of biopolitics, which is far from being a closed lexis, and some of the theoretical instruments and perspectives offered by Foucault can substantiate a diagnostic relation to the current social reality in which the human body and the various aspects of human life are constantly targeted by political and social strategies and are subject to heated debates.

References

Agamben, Giorgo. 1998. *Homo sacer: Sovereign power and bare life*. Trans. D. Heller-Roazen. Stanford: Stanford University Press.

Bourdieu, Pierre. 1992. Identity and representation. In *Language and symbolic power*, ed. Pierre Bourdieu, 220–251. Trans. G. Raymond and M. Adamson. Cambridge: Polity Press.

Brown, Wendy. 1995. *States of injury: Power and freedom in late modernity*. Princeton: Princeton University Press.

Chen, Y. Mel. 2012. *Animacies: Biopolitics, racial mattering, and queer affect*. Durham: Duke University Press.

Chriss, James. 2007. *Social control: An introduction*. London: Polity Press.

Cohen, Ed. 2009. *A body worth defending: Immunity, biopolitics, and the apotheosis of the modern body*. Durham: Duke University Press.

Cooper, Davina. 1995. *Power in struggle: Feminism, sexuality, and the state*. New York: New York University Press.

Esposito, Roberto. 2008. *Bíos: Biopolitics and philosophy*. Trans. T. Campbell. Minneapolis/London: Minnesota University Press.

Esposito, Robert. 2011. *Immunitas: The protection and negation of life*. Trans. Z. Hanafi. Cambridge, MA: Polity Press.

Fehér, Ferenc, and Ágnes Heller. 1994. *Biopolitics*. Aldershot: Avebury.

Foucault, Michel. 1977. *Discipline and punish: The birth of the prison*. Trans. A. Sherida. New York: Random House.

———. 1981. *The history of sexuality: An introduction*. Trans. R. Hurley. London: Penguin.

———. 2003. *"Society must be defended". Lectures at the Collège de France 1975–76*. Trans. D. Macey. New York: Picador.

———. 2007. *Security, territory, population: Lectures at the Collège de France 1977–78*. Trans. G. Bucknell. London: Palgrave Macmillan.

———. 2008. *The birth of biopolitics: Lectures at the Collège de France 1978–1979*. Trans. G. Bucknell. London: Palgrave Macmillan.

Hardt, Michael, and Antonio Negri. 2000. *Empire*. Cambridge, MA: Harvard University Press.
———. 2009. *Commonwealth*. Cambridge, MA: Harvard University Press.
Heyes, Cressida J. 2007. *Self-transformations: Foucault, ethics and normalized bodies*. Oxford: Oxford University Press.
Lemke, Thomas. 2011. *Biopolitics: An Advanced Introduction*. Trans. E.F. Trump. New York/London: New York University Press.
Liesen, Laurette, and Mary Barbara Walsh. 2012. The competing meanings of "biopolitics" in political science: Biological and postmodern approaches to politics. *Politics and the Life Sciences* 31 (1–2): 2–15.
Losoncz, Márk, and Ádám Takács. 2016. Biopolitics. In *Encyclopedia of global bioethics*, ed. Henk ten Have, 372–380. Dordrecht: Springer.
Nealon, Jeffrey T. 2008. *Foucault beyond Foucault. Power and its intensification since 1984*. Stanford: Stanford University Press.
Oulc'hen, Hervé, ed. 2014. *Usages de Foucault*. Paris: PUF.
Puar, Jasbir K. 2007. *Terrorist assemblages: Homonationalism in queer times*. Durham: Duke University Press.
Rabinow, Paul, and Nikolas Rose. 2006. Biopower today. *BioSocieties* 1: 195–217.
Riley, Sarah. 2008. *Critical bodies: Representations, practices and identities of weight and body management*. London: Palgrave Macmillan.
Rodger, John. 2008. *Criminalising social-policy: Anti-social behaviour and welfare in de-civilised society*. Devon: Willan Publishing.
Rose, Nikolas. 1996. *Inventing our selves: Psychology, power and personhood*. Cambridge: Cambridge University Press.
———. 2007. *The politics of life itself: Biomedicine, power, and subjectivity in the twenty-first century*. Princeton: Princeton University Press.
Spade, Dean. 2011. *Normal life: Administrative violence, critical trans politics, and the limits of law*. Cambridge, MA: South End Press.
Stoler, Ann Laura. 1995. *Race and the education of desire. Foucault's history of sexuality and the colonial order of things*. Durham: Duke University Press.
Synnott, Anthony. 1993. *The body social: Symbolism, self and society*. London: Routledge.
Takács, Ádám. 2004. Between theory and history: On the interdisciplinary practice in Michel Foucault's work. *Modern Language Notes* 119 (4, French Issue (September, 2004)): 869–884.
Taubler, Alfred I. 2005. *Patient autonomy and the ethics of responsibility*. Cambridge: The MIT Press.
Ten Have, Henk, Gerrit Kimsma, and Stuart Spicker, eds. 1990. *The growths of medical knowledge*. Dordrecht/Boston/London: Kluwer.
Turner, Bryan S. 1992. *Regulating bodies: Essays in medical sociology*. London: Routledge.

Part II
Applications: Contemporary Biopolitics and Bioethical Issues

Chapter 2
From Biopower to Empower – How to Get Plump, or Why Do We Choose What We Choose?

Ignaas Devisch

Abstract With concepts like 'governmentality', 'pastoral power' or 'population', in his later lectures on biopolitics, Michel Foucault attempted to conceive the question how the care for public health became indeed a central task and for politics and for all of us. Health is not only on the political agenda of many governments (governmentality), it's presence runs as an bioimperative through the whole of society. It appeals to all of us (a totalizing technique) and to each one of us (individualized). There is obviously something wrong with us (pastoral power) if we don't obey the imperative. I apply his analysis to the contemporary discourse on health promotion, the growing interest of the government and insurance companies in our daily activities and their attempt to interfere in it. Analysing the case of obesity, I will explore how the discourse on patient empowerment is actually the ethico-medical way through which people are governed. Far from a neutral plead, patient empowerment puts the individual responsibility for our health right at the centre of today's medical discourse. Being unhealthy has become the synonym for not having done enough. It is therefore no coincidence that today, public health, especially in industrialised countries, has also become a question of having no longer access to the health insurance due to 'bad behaviour', of being excluded from health facilities, of food industry trying to get a grip on our food habits and tastes with food supplements, etcetera. What we are dealing with today – public health as an explicit task of contemporary politics – not only can be understood as the culmination of an ongoing process of government of our daily life out of a medical perspective; it is also an explicit political evolution which needs to be made explicit, in order to understand the biopolitical ideology behind it.

Keywords Foucault • Biopower • Governmentality • Obesity • Public health • Risk

---

I. Devisch (✉)
Ghent University, De Pintelaan 185, 6K3 Vakgroep Huisartsengeneeskunde,
9000 Gent, Belgium
e-mail: Ignaas.Devisch@ugent.be

P. Kakuk (ed.), *Bioethics and Biopolitics*, Advancing Global Bioethics 8,
DOI 10.1007/978-3-319-66249-7_2

2.1 Introduction

For a long time, being fat or corpulent has been a symbol of high social status, and sometimes, layman people did everything they could not to look skinny. At the end of nineteenth century for instance, people could buy products which promised them to get plump. To mention only one example: a product called 'fatten U food' advertised with the slogan 'don't look like the poor unfortunate'. (http://wholehealthsource.blogspot.sk/2011/04/fat-ten-u.html).

Indeed, it is true what Norbert Elias described in *The process of civilization*: lower social classes obtain a mimicry of higher social classes, so they can pretend as if they actually are high class people (Elias 1995). Once this is the case, the higher social classes need to differ themselves from the lower ones. And indeed, if we look at the social status of weight, this is the case. Compared with nineteenth century, things have turned upside down. While the poor have gained weight, the corpulent director of a factory has become a CEO running marathons. For those not convinced yet, as always the market helps them a little bit: for years now, we can find advertisements in health magazines who talk about wonder diets, this time not to get plump, but to lose weight.

Now, what has happened in between today and the advertisement? Some have tried to summarize this evolution with the phrase 'from fatness to fitness' (Zanker and Gard 2008). Being corpulent is no longer perceived as a sign of wealth – like: look at me, I can eat more than I have to – but rather it has become a marker of laziness, lack of character or of not being successful in life in general.

Of course, I am very aware that today, we are being confronted with a worldwide increase of obesity, sometimes called 'globesity', and a part of the evolution from fatness to fitness has to be explained starting from this fact (http://www.who.int/nutrition/topics/obesity/en/). Globesity is an immense problem that is still increasing (Alemanno and Carreno 2011; Davey and Stanton 2004). I presume we are all aware of that but this is not what I want to talk about in the first place. To understand why we have evolved from fatness to fitness, we also need an analysis of the shifts in policy discourse on weight and our so called 'individual responsibility'. This analysis includes a short overview of the changes political power has undergone through this evolution.

2.2 Three Kinds of Power Relationships

Of course, we must not forget that before the rise of health politics and biopower, another kind of power was at work all over European countries, what is called sovereign power in political theory. *The Prince* sketched by Niccoló Machiavelli is one of the most classic examples of this idea of power (Machiavelli 1981). The Prince's main occupation is how to maintain or even augment the territory of his country and to protect its citizens. He is not occupied with the interests of his subjects, but with his territory and how to defend it. The motto of the sovereign power towards the citizens is clear and straightforward: you have to. Central to sovereign power is the difference between the legislator and the citizens. Citizens have to obey what the legislator has in mind.

This traditional function of (sovereign) power, Foucault explains in his seminars on biopolitics by the end of the seventies, is being put under pressure from the sixteenth century on and finally shifts during the eighteenth century with the rise of health politics and the welfare state. While Machiavelli speaks of sovereignty, territory and subjects, for example Guillaume de la Perrière writes in his *Le miroir politique* about government as 'the right disposition of things' (Foucault 2007, p. 92). For La Perrière and many others of this period in western Europe, government is no longer a question of territory but of people in relation to their goods, environment and conditions of life. Foucault stresses the fact that with the rise of this 'new art of governing', not only has the area of governing shifted, but also the way power is understood.

As many of you know, Foucault calls this new way of power relationships in society, governmentality (Burchell et al. 1991). Governmentality is characterized by a series of techniques, procedures, institutions or calculations by which the state tries to manage the population. In Foucault's jargon of his earlier work, this is a matter of being disciplined or of biopower. One by one, Foucault developed these theoretical concepts to explain how our lives (our 'bios') are increasingly captured by 'dispositives' or 'micromechanisms of power' determining our behaviour and self-understanding (Bishop and Jotterand 2006; Ciccarelli 2008; Holmes 2002).

Until he died, Foucault kept fine-tuning his conceptual apparatus to map this shift from sovereign power to health politics and the rise of the modern welfare state. In a comment on Foucault, Louis McNay rightly accentuates the fact that: "The notion of government does not replace the theory of disciplinary biopower; rather modern societies are characterized by a triangular power complex: sovereignty-discipline-government or 'governmentality'" (McNay 1994, p. 117).

Different from sovereign power, governmentality and biopower are more about 'I have to'. Not because some Uncle Sam is pointing his finger at us, but because citizens have interiorised the way they should behave, and thereby, we have learned to discipline ourselves. Due to the collapse of traditional sovereignty, Foucault stresses, power is present in a far more subtle way and therefore rather difficult to localize.

While Foucault described the rise of health politics and the development of governmentality in the eighteenth century, I think today we are confronted with a new shift in medicine, which, in its turn, gives birth to a new kind of power relationships which I call empower. For years now, empowerment is a key term in many discussions on healthy lifestyle and personal responsibility, or in many political debates on health, poverty and education. Apparently, today we all have to be empowered: let them be children, elderly, disabled or simply patients, we must be empowered (Erhart et al. 2008; Gagnon et al. 2006; Kligler 2007; Loukanova and Bridges 2008).

In the medical world, in all its variety, empowerment stems from the idea that medical professionals should no longer paternalize patients. Instead, patients are considered to manage their medical condition (medication, diet prescriptions, using medical devices) or should be trained towards 'active and responsible agents. Far from a neutral evolution, the goal of empowerment is more 'participation' and 'self-management of individuals' (Mola et al. 2008; Nirschl 1999; Salmon and Hall 2003; Santiago et al. 2003). I will come back to this.

Since the rise and proliferation of predictive medicine, the stage of healthcare has changed in the most profound way (Silvers 2007; Tam 2008; Wolf and Kahn 2007). The possibility to predict diseases by screening individuals on certain parameters, or by unraveling their genetic structure in order to map the future risks people are running, is a major shift in the way health and sickness are related to each other. The ability to predict diseases and guard us against them, is an enormous benefit of current medicine (Johnson and Millstein 2003). This, however, is not the question.

The possibility of prediction of risks and diseases has substantial implications for the understanding of our health condition or identity in general. The moment one can predict future health risks, the anticipation of these risks in another lifestyle becomes central (Huibers and Spijker 1998). As the risks are being individualised, from then on, it is up to the individual to make use of the available opportunities as much as possible. Of course, this can have positive effects but if the prevention from getting sick becomes a main destiny in life, being healthy becomes an never ending work in progress. Since one has never done enough to get healthy because one can never be healthy enough, there is always more to do. If things can always be better, one can always do more. Health becomes an illustration of 'Yes, I can'.

What we are dealing with is a double sided evolution. On the one hand, mainstream healthcare discourse today has shifted away from paternalism and punishment, towards a more soft and human focus on individual autonomy, empowerment and free choice. Reports or flyers stemming from governments or health insurances are presenting their health policy as if health is nothing but our free choice, opportunities and challenges. We are only invited to take pills to control your anxiety, as we are seduced to undergo a total body scan in order to assure yourself there is really nothing wrong with you.

But while society is offering us a battery of opportunities and challenges to obtain a better health, a perfidious logic is at work here: side by side with the focus on empowerment, the popularity of the idea of participation and individual responsibility has increased substantially. We are not forced to be healthy or fit, but nevertheless, it is up to us to make use of the opportunities and challenges, and it becomes our individual responsibility if we didn't meet them. What used to be an obligation, is now presented as a chance or an opportunity an individual can take, but there is still a lot of power at play in this context. Empower only seems to be of a free zone of voluntary, individual choices, but it never is.

Probably the most obvious way to understand this shift is to look at the way weight and fitness are promoted today. And because I don't want you to accuse me of mere abstract talking, I will present you briefly one case of this kind of health discourse which is so generously dispersed in our societies.

2.3 Fit and Healthy

I screened several years of information magazines of health insurances in Belgium and found multiple interesting illustrations. Though particular, these documents demonstrate what a new kind of power relationships.

I will quote from one out of many magazines. The (Dutch) number is entitled 'Health is risk management', from 2007 (http://www.mloz.be/cms/Mloz/FaxMedicaNL_3.pdf). It focuses on the question how to make of prevention a more central endeavor in healthcare, and how to influence on people's behavior. Not for one moment there is referred to (financial) punishment; 'good care' is put forward as the central objective, empowerment as the means to attain it. All the time, good care is intertwined with the empowerment of individual responsibility for our lifestyle. Each one of us is invited to change our lifestyle and is warned for the possible health consequences or our risk or 'bad' behavior. Next to that, the need of information about the personal lifestyle in daily life is explicitly mentioned: people should do more medical screenings and general check-ups, etc. When predictive medicine is presented in detail, one can read:

> We would need medical databases with biomarkers, collected from individuals whose lifestyles, eating habits and possible presence of toxic elements in the body, are screened at forehand. As these data should be cross checked with other registers such as cancer databases, we would need an amendment of the privacy law. (http://www.mloz.be/cms/Mloz/FaxMedicaNL_3.pdf, p. 17; my translation)

And further on, in the same magazine, one can read:

> Of course, the ultimate goal of prevention is to guarantee the welfare of the citizens, to prevent people dying at young age and to increase their life expectancy and quality of life. And of course, also economic considerations play a role in this process. (http://www.mloz.be/cms/Mloz/FaxMedicaNL_3.pdf, p. 17; my translation)

In more than one way, this is an interesting case. The 'of course' is presented as supporting the welfare of the citizens but in the meantime an amendment of the privacy law is necessary: apparently this good care is willing to take care of us, but it can only be realized if we say yes to a voluntary screening and check-up and to a violation of our private life.

Not by accident, the website of the insurance company appeals to companies and brands with the campaign 'Fit and healthy', and convinces CEO's of the need of a good health policy for their employees: "Go for fit and motivated employees! Make of your company a healthy one!" Given the economical perspective, the financial profit of fit employees is explained in detail and the profit of selfmanagement of health behavior is sold as the 'return on investment' of this 'good care' of the company (http://www.oz501.be/web/guest/fit-en-gezond).

As many other companies in Belgium and elsewhere, this insurance company relates the individual responsibility for our health to the general need of employees in a good condition. Individuals therefore are insisted to make of their behavior an affair of selfmanagement and selfgovernment, as their contribution to a healthy society.

This discourse is revealing because it demonstrates how a new discourse and new power relationships go hand in hand. It has substantial consequences to the way people are considered to take care of their health in person. This care presents itself as a mere positive gesture, but in the meantime, behind the curtains, negativity and punishment are waiting around the corner. It is a bit analogue to what Adorno and Horkheimer described in their *Dialektik der Aufklärung*: negativity is the inevitable flipside of a culture who aims at mere positivism (Horkheimer and Adorno 1996).

This double sided way of offering opportunities and empower people in a preferable direction is the reason why I describe this new social construct of power relationships as empower. Maybe empower does not differ in principle from biopower, but it makes a shift in the way people are appealed to in order to take care of their health. Empower is less about 'I have to' but more about 'Yes, I can'. It is a 'productive' power as Judith Butler says (Butler 2002), which aims at the creation of a certain type of citizen who is able of selfmanagement and selfgovernment (Lynch 2004; Mola et al. 2008; Moran et al. 2008), and who is not even suffering from it; the empowered individual is happy with the way he was able to manage and schedule his goals and objectives. The empowered individual is not making protest against his subjection to a law, but he is enjoying his well-trained ability to construct his own identity 'all by himself'. While the world of biopower, as the German philosopher Byung-Chul Han argues, is crowded with *negativity* (prisons, hospitals, institutes); today, we start from an excess of *positivity*: we are all challenged by opportunities, individual merits and high potentials (Han 2011). The motto of empower therefore, is 'yes I can'. 'Who is am' is considered to be the result of my personal merits.

2.4 From Paternalism to Empowerment

In a preliminary way, I will sketch some characteristics of the paradigm of empower. The list of characteristics is far from exhaustive, but it's better to dare and think of new concepts, than simply repeat the old ones.

First of all, the explicitly invitational character of this discourse is remarkable. The single case of the insurance company illustrates the company avoids any reference to punishment or to tell people how to behave; instead, the need of self-management is put forward as a central aspect of good care. The discourse is about 'we want to invite you', 'we want to support you'. This is not a single case: hospitals, ministries of health, health departments, they cannot stop talking about the importance of the empowerment of the patient. Considering that the importance of patient autonomy has increased, health professionals are no longer supposed to paternalize patients – 'you have to live a healthy lifestyle because I tell you to do so' – but 'empower' them so they would be able to make the healthy choices by themselves (http://www.oz501.be/web/guest/fit-en-gezond).

But how free are we actually? While the objective remains the same as it used to be – a healthy life – the method has changed. Patients are supposed to be empowered and to be capable of self-management, but who sets the norms the empowerment is headed towards? Though the discourse on empowerment is about 'autonomy and choice', the choices we have to make are outlined in advance. Next to that, a certain type of subjectivity is put forward as 'normal' or 'good' and those individuals who defy from it, are subjected to strategies of risk management. If people are empowered to 'help themselves', the empowerment is always pointed at some objective. In setting up this objective power is definitely at work (http://www.oz501.be/web/guest/fit-en-gezond).

As Gilbert has demonstrated in an analysis of empowerment in nursing practice, though reflective practice and clinical supervision are progressively asserting hegemony upon nursing practice with claims of emancipation and empowerment, this empowerment is being achieved in an environment where there is little critical debate about the assumptions on which these practices are based (Gilbert 2001).

Far from a value free supportive practice, empowerment is all about "the promotion of an ethic of the self, which incites individuals to be self-managing, producing particular forms of subjectivity and modes of subjectivation", Gilbert writes (2001). Though individuals are only 'invited' to take care of their health, the norms of behavior are still set by programs stemming from health professionals or health policy makers. In particular individuals with 'risky behavior' are supposed to be empowered because they make the wrong choices according to the norms set by healthcare. Risk behavior is something to be governed, just as unemployment or lifelong learning.

The discourse of empowerment in a healthcare context therefore, is being enrolled in a quite ambiguous logic. The support for individual autonomy is advocated as a means to empower people and change their health behavior. We are empowered to make our choices, but the kind of choices we make, are set up in advance. As a consequence, we are 'free to make the right choices', or as Nikolas Rose puts it, or we are "obliged to be free" (Rose 1996). This is the second characteristic, a double bind logic of freedom and obligation.

A third characteristic is about the new political context in which empower is embedded in. In several countries such as the Netherlands, the welfare state is explicitly left behind and substituted by what is called the participation society. In a speech on national television, in 2013, King Willem-Alexander delivered a message to the Dutch people from the government: the welfare state of the twentieth century is gone. In its place a "participation society" is emerging, in which people must take responsibility for their own future and create their own social and financial safety nets, with less help from the national government. "The shift to a 'participation society' is especially visible in social security and long-term care", the king said, reading out to lawmakers a speech written for him by the Dutch Prime Minister (http://www.rijksoverheid.nl/documenten-en-publicaties/toespraken/2013/09/17/troonrede-2013.html).

Anyway, what King Willem Alexander was talking about, is nothing but the third characteristic I am putting forward: participation instead of care. Interestingly, the king also said that nowadays, people expect and "want to make their own choices, to arrange their own lives, and take care of each other." (http://www.rijksoverheid.nl/documenten-en-publicaties/toespraken/2013/09/17/troonrede-2013.html).

As the 'participation society' has been on its way for some time and benefits such as unemployment compensation and subsidies on health care have been regularly pruned for the past decade, this brings me to the fourth and last characteristic, the becoming conditional of certain rights and facilities.

The more life and health are presented as the result of our free and independent choices, the more we have to deserve our civil rights, it seems. If health is, as I sketched above, an ongoing work in progress, not only is it up to ourselves to get along with the invitations and opportunities, we are also confronted with the fact that we have to 'merit' certain facilities or rights. In origin, many social security systems were based upon the idea of solidarity, no matter how people behaved in their personal life. 'Merit' sets this idea upside down (Verhaeghe 2014).

While the idea of security is left behind and substituted by risk, a lot of polls have been executed in many countries to discuss the idea of solidarity (Ter Meulen and Jotterand 2008; Ter Meulen and Maarse 2008). People would wonder if smokers, drinkers or simply eaters should be able to enjoy health facilities. Often, rights or facilities have become conditional: individuals are only supported when they meet certain conditions. In many cases this evolution is quite analogue: first people are supported and rewarded if they do something extra like visiting fitness clubs; several insurance companies even contribute in the consumption of cholesterol free butter. Once this practice is settled, slowly but steadily, people are no longer rewarded for what they do, but punished for what they didn't do. No one is obliged to go out running, but when asking for a health insurance, one is punished for having a bad cholesterol.

As a consequence, the logic of empowerment is a perfidious one: it presents itself as supportive and overtly positive because people are free to make their own choices; but underneath this liberty, a moral path of good and bad choices is being developed and people are empowered only to make the 'good choices'. Being empowered to take care of our health, is at least an ambivalent entrepreneurship. We are free to choose, but obviously, we are governed or empowered which choices to make. In the era of empower, paternalism has not been abandoned, but it is integrated into a discourse of 'participation'.

Nevertheless, the discourse on empowerment leans on a kind of voluntarism when it comes down to our personal choices. A healthy lifestyle is often put forward as a deliberate choice of free individuals, and the consequence of our will and character. If we really want it, it will happen. Who we are, would be the result of our rational choices. As a consequence, citizens are supposed to consider their choices in the light of a healthier lifestyle, but because they don't do it yet, they have to be empowered (Devisch and Deveugele 2010).

2.5 Choice and Circumstances

This brings me to the last section of my paper and, finally, to the explication of my cryptic subtitle 'why do we choose what we choose'. I want to answer this question by sketching a topical example, obesity. As lifestyle more and more becomes a crucial factor in the treatment of obese patients, lifestyle is already an issue in current policy discussions about obesity. A crucial point in this discussion deals with the way we understand the concept of choices and free will: is it fair to hold obese patients fully responsible for their condition, because it is simply the result of their deliberate lifestyle choices?

Although at first glance our personal choices seem simple and unambiguous, the reality behind them is complex. Food and physical activity are more than aspects of our health. They are an integral part of our sense of well-being as persons, individually as well as socially, and therefore cover many important values of our lives which are related to our food patterns, social behavior and aspects of socialization. It is common sense to believe that food choices are strictly personal, but do we, as individuals, really choose what we choose? There are many reasons to doubt this.

Let me take a closer look at this: obesity has a direct relationship with eating, but it is related to so much more. Of course, individual access to education about healthy food, and stimulating a healthy diet together with physical activity in those who are overweight, is the main pathway to prevent or overcome obesity. But there are several mediating influences, to name the most important ones: genetic predisposition, behavioral and psychological characteristics, parenting and education, economic and social factors (Fahlquist 2006; Murray 2008; Tanumihardjo et al. 2007; Wyatt et al. 2006).

As we learned from the advertisement at the beginning, everyone lives in context. It is of course the individual who makes his or her own choices, but the context within which he or she can choose is not a matter of choice. This social and economic context is more than just a footnote in the debate on lifestyle; it is a crucial element. Obesity is especially a problem of contemporary society. This is not simply because we are weak or unwilling to practice sports or to undertake any physical activity. Various structural facts direct our behavior and lifestyle.

The key point here is not that we are devoid of responsibility for our own choices. Of course, the individual is responsible, but the question is what makes us choose what we choose? (Devisch and Deveugele 2010) Or to put it bluntly: are our food choices well informed and deliberate choices, and the result of mere individual lifestyle decisions? Besides differences in taste and the complex matter of how we develop taste, the question remains whether the choice of what is on our plate is only our own? As Eli Feiring writes: "It is hard to identify any action that is not partly determined by circumstance understood as the social contexts in which the individual finds herself or her traits of character (included the ability to choose)" (Feiring 2008).

If we conceive of lifestyle as a deliberate and rational choice of an independent individual, and consequently view obesity as the result of this, then the question remains why are so many people acting in the same way at the same time and place? Is this really because all of a sudden, one by one, half of the western population quite independently chose to eat and drink too much and/or to no longer have any physical activity? (Devisch and Deveugele 2010).

Conversely, I am not opting for the former "Marxist" thesis by saying that the "system" is responsible for everything. Instead, I want to suggest a counterintuitive thesis: there are aspects of our behavior for which no one is responsible. By this I mean that much of our choices are developed as a strategy for living our lives or, analogously to what Carl Elliott writes about values in his *A Philosophical Disease*, "the result of cultural factors beyond our reach" (Elliott 1999). As long as lifestyle is considered to be based solely on independent, conscious and deliberate choices, be they taken individually or collectively, we will never solve the question of why people are "choosing" one thing or refusing another. We don't choose a lot of our choices.

2.6 Conclusions

The shift from biopower to empower is a subtle but an important one if it comes down to our health. If 'yes, I can' is our motto, then not having done enough while you could have done more, is worse than not having fulfilled your duty; but most of all, it presumes that everything we do, is the result of individual, deliberative choices.

Although we make our own choices, our health is far more than the personal expression of who we are; it is also the result of cultural or societal habits, patterns or evolutions far beyond our individual reach. Given the subtle powerful influences on our value systems and choices, it should not surprise us that we present ourselves as, or aspiring, sportive, healthy and responsible people. The "self-steering mechanisms of individuals" are the result of widespread and ongoing strategies of empowerment towards a healthier lifestyle. Therefore today, everyone is looking at the diets of the second advertisement, because the fatten U food is everywhere.

As Nikolas Rose summarizes in *Governing the Soul*: "The political subject is now less a social citizen with powers and obligations deriving from membership of a collective body, than an individual whose citizenship is to be manifested through the free exercise of personal choice among a variety of marketed options." (Rose 1990) Empower hopefully sheds a new light on what health discourse is about today: obliging people to be free.

Acknowledgements and Conflicts of Interest This invited article has drawn indirectly and in the last section verbatim on the content of the previously published paper 'FOUCAULT AT THE BEDSIDE: A CRITICAL ANALYSIS OF EMPOWERING A HEALTHY LIFESTYLE '((2015) Devisch, Ignaas and Vanheule Stijn, Foucault at the bedside: a critical analysis of empowering a healthy lifestyle. Journal of Evaluation in Clinical Practice. Doi: 10.111/jep.12329). This derivation has been utilized to re-present and emphasize previously articulated observations, opinions and contentions. The author of the current paper declares no conflicts of interest.

References

Alemanno, A., and I. Carreno. 2011. Fat taxes in the European Union between fiscal austerity and the fight against obesity. *European Journal of Risk Regulation* 4.

Bishop, J.P., and F. Jotterand. 2006. Bioethics as biopolitics. *The Journal of Medicine and Philosophy* 31 (3): 205–213.

Burchell, G., C. Gordon, and Peter Miller. 1991. *The Foucault effect. Studies in governmentality with two lectures by and an interview with Michel Foucault*. Hertfordshire: Harvester Wheatsheaf.

Butler, J. 2002. What is critique? On Foucault's virtue. In *The political*, ed. D. Ingram. Oxford: Blackwell.

Ciccarelli, R. 2008. Reframing political freedom in the analytics of governmentality. *Law and Critique* 19 (3): 307–327.

Davey, R.C., and R. Stanton. 2004. The obesity epidemic: Too much food for thought? *British Journal of Sports Medicine* 38 (3): 360–363.

Devisch, I., and M. Deveugele. 2010. Lifestyle: Bioethics at a critical juncture. *Cambridge Quarterly of Healthcare Ethics* 19 (4): 550–558. doi:10.1017/s0963180110000502.

Elias, N. 1995. *Über den Prozess der Zivilisation*. Frankfurt am Main: Suhrkamp.

Elliott, C. 1999. *A philosophical disease: Bioethics, culture and identity*. London/New York: Routledge.

Erhart, M., N. Wille, and U. Ravens-Sieberer. 2008. Empowerment of children and adolescents – The role of personal and social resources and personal autonomy for subjective health. *Gesundheitswesen* 70 (12): 721–729. doi:10.1055/s-0028-1103261.

Fahlquist, J.N. 2006. Responsibility ascriptions and public health problems. *Journal of Public Health* 14 (1): 15–19.

Feiring, E. 2008. Lifestyle, responsibility and justice. *Journal of Medical Ethics* 34 (1): 33–36.

Foucault, M. 2007. *Security, territory, population: Lectures at the Collège de France, 1977–1978*. New York: Palgrave Macmillan.

Gagnon, M., R. Hibert, M. Dube, and M.F. Dubois. 2006. Development and validation of an instrument measuring individual empowerment in relation to personal health care: The Health Care Empowerment Questionnaire (HCEQ). *American Journal of Health Promotion* 20 (6): 429–435.

Gilbert, T. 2001. Reflective practice and clinical supervision: Meticulous rituals of the confessional. *Journal of Advanced Nursing* 36 (2): 199–205.

Han, B.-C. 2011. *Topologie der Gewalt*. Berlin: Matthes & Seitz.

Holmes, D. 2002. Police and pastoral power: Governmentality and correctional forensic psychiatric nursing. *Nursing Inquiry* 9 (2): 84–92.

Horkheimer, M., and T. Adorno. 1996. *Dialektik der Aufklärung. Philosophische fragmente*. Frankfurt am Main: Fischer Wissenschaft.

Huibers, A.K., and A. Spijker. 1998. The autonomy paradox: Predictive genetic testing and autonomy: Three essential problems. *Patient Education and Counseling* 35 (1): 53–62.

Johnson, S.B., and S.G. Millstein. 2003. Prevention opportunities in health care settings. *The American Psychologist* 58 (6–7): 475–481.

Kligler, B. 2007. Patient empowerment: Taking integrative medicine to the next level. *Journal of Alternative and Complementary Medicine* 13 (2): 189–190. doi:10.1089/acm.2006.6343.

Loukanova, S.N., and J.F.P. Bridges. 2008. Empowerment in medicine: An analysis of publication trends 1980–2005. *Central European Journal of Medicine* 3 (1): 105–110. doi:10.2478/s11536-007-0066-7.

Lynch, J. 2004. Foucault on targets. *Journal of Health Organization and Management* 18 (2–3): 128–135.

Machiavelli, N. 1981. *The prince*. Harmondworth: Penguin Books.

McNay, L. 1994. *Foucault. A critical introduction*. Cambridge: Polity Press.

Mola, E., J.A. De Bonis, and R. Giancane. 2008. Integrating patient empowerment as an essential characteristic of the discipline of general practice/family medicine. *The European Journal of General Practice* 14 (2): 89–94. doi:903051824 [pii]10.1080/13814780802423463.
Moran, J., H. Bekker, and G. Latchford. 2008. Everyday use of patient-centred, motivational techniques in routine consultations between doctors and patients with diabetes. *Patient Education and Counseling* 73 (2): 224–231.
Murray, S. 2008. Pathologizing "fatness": Medical authority and popular culture. *Sociology of Sport Journal* 25 (1): 7–21.
Nirschl, R.P. 1999. Patient empowerment: Individual control of health benefits. *Orthopedics* 22 (8): 727–728.
Rose, N. 1990. *Governing the soul: The shaping of the private self*. London: Routledge and Kegan Paul.
———. 1996. The death of the social? Re-figuring the territory of the government. *Economy and Society* 25 (3): 327–356.
Salmon, P., and G.M. Hall. 2003. Patient empowerment and control: A psychological discourse in the service of medicine. *Social Science & Medicine* 57 (10): 1969–1980. doi:S0277953603000637 [pii].
Santiago, L.E., C. Santiago, J. Texidor, G. Montalvo, and A.Y. Rivera. 2003. Social participation and community empowerment in public health. *Puerto Rico Health Sciences Journal* 22 (1): 39–41.
Silvers, A. 2007. Predictive genetic testing: Congruence of disability insurers' interests with the public interest. *The Journal of Law, Medicine & Ethics* 35 (s2): 52–58.
Tam, P.K.H. 2008. Towards predictive, preventive, and personalized paediatric surgery. *Journal of Pediatric Surgery* 43 (2): 267–273.
Tanumihardjo, S.A., C. Anderson, M. Kaufer-Horwitz, L. Bode, N.J. Emenaker, A.M. Haqq, et al. 2007. Poverty, obesity, and malnutrition: An international perspective recognizing the paradox. *Journal of the American Dietetic Association* 107 (11): 1966–1972.
Ter Meulen, R., and F. Jotterand. 2008. Individual responsibility and solidarity in European health care: Further down the road to two-tier system of health care. *The Journal of Medicine and Philosophy* 33 (3): 191–197. doi:jhn012 [pii]10.1093/jmp/jhn012.
Ter Meulen, R., and H. Maarse. 2008. Increasing individual responsibility in Dutch health care: Is solidarity losing ground? *The Journal of Medicine and Philosophy* 33 (3): 262–279. doi:jhn011 [pii]10.1093/jmp/jhn011.
Verhaeghe, P. 2014. *What about me?: The struggle for identity in a market-based society*. Brunswick: Scribe.
Wolf, S.M., and J.P. Kahn. 2007. Genetic testing and the future of disability insurance: Ethics, law & policy. *The Journal of Law, Medicine & Ethics* 35 (s2): 6–32.
Wyatt, S., K. Winters, and P. Dubbert. 2006. Overweight and obesity: Prevalence, consequences, and causes of a growing public health problem. *American Journal of the Medical Sciences* 331 (4): 166–174.
Zanker, C., and M. Gard. 2008. Fatness, fitness, and the moral universe of sport and physical activity. *Sociology of Sport Journal* 25 (1): 48–65.

Chapter 3
Biological or Democratic Citizenship

Vilhjálmur Árnason

Abstract This chapter starts with a description of an Icelandic case of a national collection of biosamples in the spring of 2014. A critique of the collection and a reaction to that critique is presented. In the wake of these descriptions, the case and its relation to population genetic research is discussed in light of two different views on scientific citizenship. The first view is Rose's and Novas' analysis of "biological citizenship" (Rose N, Novas C. Biological citizenship. In: Ong A, Collier SJ (eds) Global assemblages: technology, politics, and ethics as anthropological problems. Blackwell Publishing, Oxford, pp 439–463, 2004) and the other view draws on deliberative democratic theory (I have discussed these views in Árnason 2013 and I draw largely on that paper in my theoretical discussion here). It is shown how these two views disclose very different features of the collection of biosamples and the population genetic project in Iceland. It is argued that these two views provide complementary perspectives that are both necessary to throw light on the bioethical and biopolitical issues relevant for analyzing this case.

Keywords Biological citizenship • Rose • Novas • Biosamples • Biovalue • Deliberation

3.1 A Case From Iceland

In early May 2014 the genetic research company deCODE Genetics launched a massive collection of biosamples in Iceland under the banner of "Urgent Call in the Service of Science". The collection was performed in co-operation with the Icelandic National Rescue Team whose members collected the samples. The National Rescue Team was to receive a certain amount (approximately 15 USD or 13.5 euros) from deCODE Genetics for each biosample collected. The company posted a package to over 100.000 addressees in Iceland, containing consent form with information about the intended research on the samples, a stick for a collection of the sample from the

V. Árnason (✉)
Faculty of History and Philosophy, University of Iceland, Gimli, Sæmundargata 2, IS 101 Reykjavik, Iceland
e-mail: vilhjarn@hi.is

P. Kakuk (ed.), *Bioethics and Biopolitics*, Advancing Global Bioethics 8,
DOI 10.1007/978-3-319-66249-7_3

participant's mouth. The participants were expected to pack the sample and hand it over to a member from the National Rescue Team who would pay them a visit.

This national collection of biosamples was launched suddenly by a well-designed PR campaign where scientists, academics, politicians and artists joined hands in appealing to the nation to jump on the bandwagon for a good cause. Among prominent people who appeared in colourful advertisements were the minister of health and welfare, the mayor of Reykjavik, the dean of the school of health sciences at the University of Iceland, the dean of the medical faculty of the same university and some popular Icelandic musicians. Advertisements also showed medical staff with members of the National Rescue Team (picture) with the logo of deCODE Genetics in the background.

3.2 Critique of the Campaign

A small group of academics responded to the campaign with some critical remarks which were sent to the Icelandic media. These remarks were signed by nine people, most of whom are associated with The Centre for Ethics at the University of Iceland. The main points of criticism of the campaign were the following[1]:

[1] The criticism appeared in two short media announcements. The first, dated May 9 2014, appeared widely in the media under the title "Gagnrýna lífsýnasöfnun ÍE" (Criticism of the collection of biosamples by deCODE). This can be accessed at http://www.ruv.is/frett/gagnryna-lifsynasofnun-ie. The second, dated May 13 2014, was a response to a declaration from a group of supporters of the campaign, "Viðbrögð við yfirlýsingu vísindamanna" (Response to the declaration of scientists). This can be accessed at http://www.dv.is/frettir/2014/5/14/sidfraedingar-svara-gagnryni-laekna-tilgangurinn-helgar-ekki-medalid/

1. The co-operation of the researcher with the National Rescue Team created an undue pressure on participants to give their biosamples. The National Rescue Team is a very popular voluntary organization in Iceland. Their members often undertake heroic efforts and risk their own lives in searching for and saving people in dire circumstances, for example in the Icelandic highlands. Since the National Rescue Team was to receive a payment for each sample collected many people might find it difficult to refuse participation.
2. The group argued that the information provided in the consent form that the participants were asked to sign was misleading. The information mentioned only that the biosamples were to be collected for the purpose of having a comparative population to the previous population genetic research that had been done by the company. As in other research projects of the company, the participants had the options of signing two different types of consent. The first is a limited consent which restricts the use of the biosamples to this particular comparative research, after which it would be destroyed and not stored in the company's biobank for further research. The second is a wider consent which in addition to the first allows that the sample be stored in the biobank as part of the company's database resource for further genetic research, subject to permissions of the National Bioethics Committee and The Data Protection Authority. This "further genetic research" is in principle unforeseen and not discussed in the consent form except in very general terms. The critics point out that novel research possibilities, such as whole genome sequencing, that have often been discussed in closed professional meetings, are not mentioned in the consent form for the collection under discussion. While previous participants had mostly been selected from targeted disease groups, this campaign focused on people who had not participated in deCODE Genetics research before. The critics argued that the ulterior motive for this massive collection of biosamples was to extend the deCODE Genetics biobank as a resource for further genetic research. It should be added here, though it was not mentioned in the point of the critics of the campaign,[2] that this extension of the biobank is likely to increase the value of the company which was recently acquired by the American biotech company Amgen.
3. The campaign was launched secretly overnight, so to speak, without any public dialogue in advance. Granted that it would have been difficult to describe future genetic research in a short consent form, the critics emphasize that it would have been appropriate and necessary to prepare such a massive project by an informed public discussion about the population genetic research being undertaken by the company and planned in the future. The PR method chosen by the company aimed at creating a national herd behavior where people would be infected by the enthusiasm of contributing to a good cause under a considerable time pressure. This, the critics argued, does not befit a collection of biosamples for population research purposes; rather, the aim should be to facilitate an informed and critical

[2] There was some discussion of this in the Icelandic media. Some of the media actively supported the campaign while others engaged in critical discussion, mostly based on the critical remarks of the group associated with The Centre for Ethics at the University of Iceland.

public deliberation about the subject matter. It needs to be mentioned in this context that contrary to other Nordic countries, no national ethics council, such as the Danish Council of Ethics or the Norwegian Biotechnology Advisory Board, exists in Iceland. It is part of the mandate of such national councils to facilitate informed public debate about developments in the biomedical sciences and the moral, social and policy implications of the use of genetics and biotechnology. Efforts to ignite interest among Icelandic authorities to found such a council have not been successful. The Centre for Ethics at the University of Iceland, in co-operation with professional organizations, try to put bioethical issues on the agenda, but they do not have legal mandate to facilitate public dialogue with the aim of informing the legislator about public concerns and principled positions.

4. The critics argued that the way in which this collection of biosamples was performed could undermine trust in scientific research in the long run. They pointed out that the willingness of Icelandic citizens to participate in scientific research is a valuable resource that should be handled with care. Using PR methods which aimed at maximizing the collection of biosamples and precluding informed public debate would not be conducive to trustworthy practices.
5. Conflicts of interest. It is a worry and another threat to trustworthiness how close connections are in Iceland within the scientific and academic community. The company deCODE Genetics is comparatively a gigantic player on the Icelandic scene and it has built up a valuable research resource which cannot be accessed except through co-operation with the company which also increases opportunities for funding of research. Recently, ties have also been strengthened between the company and the University of Iceland which has boosted its ranking in international university performance tables. As a consequence, there is little resistance within the scientific and academic community to questionable projects like the national campaign discussed here.
6. To sum up, the critics argued that the method chosen for the campaign for a collection of biosamples was contrary to the spirit of research ethics and good democratic practices. The company should have shown the nation more respect by addressing people as thinking citizens who would like to be informed and participate in a public dialogue about the relevance of the project rather than aiming at herd behavior for quick gain. Even though there had been a considerable amount of discussion in relation to the Health Sector Database project around the turn of the century, that debate had been polarized and ill-informed and was in any event no substitute for a public dialogue preceding the national collection of biosamples.

The main spokesmen of the campaign "Urgent Call in the Service of Science", most of whom were medical doctors, responded harshly to these critical points.[3]

[3]The declaration of the scientists, "Yfirlýsing vísindamanna um Útkall í þágu vísinda í kjölfar yfirlýsingar siðfræðinga og nokkurra annarra fræðimanna" (Declaration of scientists about the Urgent call in the service of science in the wake of a declaration of ethicists and some other scholars) can be accessed at: http://www.mbl.is/media/30/7630.pdf. It was signed by 36 people, most of whom are practicing physicians and researchers at the National University Hospital in Reykjavik.

They stated that it was wrong to disturb and stand in the way of such a good cause. It was dangerous, they maintained, to spread seeds of suspicion about the important research that this collection served and could not be carried out without the support of the Icelandic public. They argued that Icelanders were generally well acquainted with the population genetics research carried out by deCODE Genetics and flatly rejected that there was a lack of public discussion in Icelandic society about the project. The criticism was seen as paternalistic, showing distrust in the ability of prospective participants to make up their mind in an informed way. The spokesmen of the campaign also emphasized that the project had been approved by the National Bioethics Committee, understandably, since this research project was simply a continuation of the pervasive population genetics research that had been carried out by the company for several years.[4]

3.3 The Perspective of Biological Citizenship

It is instructive to analyse this Icelandic case in light of a seminal article on biological citizenship by the sociologists Nikolas Rose and Carlos Novas.[5] Their analysis is largely framed in a discourse of production and marketing strategies. The authors show through several examples how the citizens-consumers are made and make themselves in relation to available and emerging biotechnology and genomics. They examine the relationship between genetic research and the economy. In that context they refer to Iceland where there has been heavy emphasis on population genomics. Rose and Novas write:

> Contemporary biomedicine, by rendering the depths of the body visible, intelligible, calculable and capable of intervention at a molecular level, makes it amenable to the production of economic value. In many ways, what is being accomplished through the life sciences is a kind of 'flattening' of the vital processes of the body. This not only enables these 'surfaces' to become equivalent with one another at the most basic biological level, but also allows them to be enfolded within processes of capital or social accumulation. They contain the potential to transform the vitality of each and all of us into a standing reserve for the creation of biovalue. One area where this is occurring draws on the health technologies of social citizenship and redeploys them in the service of biovalue. Two examples of this logic can be found in Sweden and Iceland.[6]

The authors have in mind the formation period of database research in Iceland where there were plans to create and operate a central Health Sector Database (HSD). The HSD was to contain health care information about Icelanders dead and alive since the beginning of medical recording, with the exception of those living individuals who opted out.[7] The discourse around this database project showed very

---

[4] Árnason (2016).

[5] Rose and Novas (2004).

[6] Rose and Novas (2004: 456).

[7] Cf. Árnason (2004).

clearly the interplay between genetic research and the production of economic value. The benefits to be reaped from this biotechnological enterprise were both medical – promising to "transform the practice of medicine"[8] by new drugs and targeted therapies – and non-medical in the form of jobs for young scientists and other economic offshoots of a thriving research company.[9] By granting deCODE Genetics an exclusive licence to create and operate HSD, Rose and Novas argue, "the state plays an active role in transforming their citizens into a potential resource for the generation of wealth and health".[10]

The plans to set up the HSD were never realized, but nevertheless deCODE Genetics has gradually built up a massive population genetic database through their research projects. Through these projects which are done in co-operation with practicing physicians, biosamples and medical information about majority of Icelanders have been collected. This information can be crossmatched with genealogical information "to identify genetic variation associated with human diseases", as stated on the company's homepage. The population genetics approach is further described:

> In our gene discovery work in Iceland, we have gathered genotypic and medical data from more than 160,000 volunteer participants, comprising well over half of the adult population. Using Iceland's uniquely comprehensive genealogical records, deCODE has also put together a genealogy database covering the entire present day population and stretching back to the founding of the country more than 1000 years ago. The combination of size of the population, the generous participation of so many people in our discovery work, the genealogies, and high quality universal healthcare make possible very large-scale studies of virtually any common disease.[11]

As mentioned here, Icelanders have been generously participating in the research projects of the company. In the terminology of Rose and Novas, they have gone willingly into the service of biovalue. Icelandic citizens have generally been open to the messages of the "promissory science"[12] undertaken by deCODE Genetics. They have invested in the economy of hope for medical and economic benefits, the production of wealth and health implied in the vision of population genomics.

Rose's and Nova's article provides in many ways a realistic and fruitful account of the actual processes of citizen-making and self-formation of the citizen-consumer. What I find more questionable in their analysis is how they relate these processes to the notion of citizenship. Rose and Novas write: "we use the term 'biological citizenship' descriptively, to encompass all those citizenship projects that have linked their conceptions of citizens to beliefs about the biological existence of human beings, as individuals, as families and lineages, as communities, as population and races, and as a species".[13] As the authors emphasize, this is a descriptive notion of citizenship which enables them to encompass various sorts of activi-

[8] Hakonarson et al. (2003).

[9] Árnason (2011).

[10] Rose and Novas (2004: 456).

[11] http://www.decode.com/research/ Accessed August 26 2015.

[12] Cf. Hedgecoe (2004).

[13] Rose and Novas (2004: 440).

ties and formation processes related to our biological existence. The reason why I find it questionable is that this descriptive use of the notion of citizenship provides no means to distinguish between passive subjugation of uninformed citizen consumers to the powers at work in the biotechnological industry, and more active and conscious projects of citizens resisting such subjugation. This is problematic because biological citizenship would then seem to be compatible with any ends and it is questionable whether it is consistent with subjugation.

We see these shortcomings of Rose's and Nova's analysis in their use of the notion of literacy which substantiates their description of biological citizenship. They take an example of how pharmaceutical companies engage in activities to educate consumers of their products which is to demonstrate "self-education of active biological citizens".[14] A good illustration of this is the authors' description of "the Prozac website" of the pharmaceutical company Eli Lilly which uses techniques "to promote a particular version of scientific or biological literacy".[15] The example shows how marketing techniques are used to intertwine advice about caring for the self with the marketing of the drug. In addition to taking the process of recovery more into their own hands, patients learn to ask their doctor to prescribe Prozac for symptoms of depression. After describing the example, the authors ask: "What kind of scientific literacy is being promoted here? What kinds of active biological citizens are being shaped, and to what ends?"[16]

These are important questions which deserve a careful discussion. Rose and Novas immediately provide a succinct answer: "This is the citizenship of brand culture, where trust in brands appears capable of supplanting trust in neutral scientific expertise."[17] Furthermore, the authors say that "this is just one example of the way in which biovalue is supplanting public value in the biological education of citizens-consumers".[18] It must be asked whether literacy in any meaningful sense is being promoted at all in these examples. According to a UNESCO position paper, literacy is the "ability to identify, understand, interpret, create, communicate, compute and use printed and written materials associated with varying contexts. Literacy involves a continuum of learning in enabling individuals to achieve their goals, to develop their knowledge and potential, and to participate fully in their community and wider society".[19] As we see from this definition, the notion of literacy has a normative critical power that is lost in Rose's and Nova's descriptive analysis which makes no distinction between what manipulates citizens and what furthers their understanding.

The same criticism applies generally to a neutral and descriptive idea of citizenship. The notion of citizenship implies not any activity of citizens. It refers to activities that are different from e.g. those characteristic of a colleague, customer or a

[14] Rose and Novas (2004: 447).

[15] Rose and Novas (2004: 448).

[16] Rose and Novas (2004: 448).

[17] Rose and Novas (2004: 448).

[18] Rose and Novas (2004: 448).

[19] Unesco Position Statement (2004).

consumer.[20] We need to be able to distinguish between what is and what is not conducive to ends of citizenship to show how citizenship can be distinguished from consumer manipulation. Otherwise it loses its distinctive dimensions. Bernard Crick and his team in Britain argued that the aim of citizenship education is to enable pupils to become "active, informed, critical and responsive citizens".[21] This is sometimes referred to as "active citizenship" but the most important aspect of it in this context is the democratic vision it implies about citizens who are motivated, informed and able to critically reflect on their society and willing to participate in processes of public deliberation about matters of common interest. Clearly, citizenship is a contested notion, but it is not helpful to use the notion of citizenship about subjugation or "subjecthood in monarchies or dictatorships"[22] because the idea of citizenship is part of a democratic ethos. The elements which enable such distinctions are erased in Rose's and Nova's account. As a consequence it lacks the emancipatory vision that accompanies classic ideas of citizenship, both liberal (rights) and republican (participation).[23]

Nevertheless, the analysis by Rose and Novas is implicitly and importantly critical by demonstrating the processes of social engineering in biopolitics. It is sceptical of citizen engagement which "transforms the vitality of each and all of us into a standing reserve for the creation of biovalue". If we return to the Icelandic case about national collection of biosamples we see how well it can fit into this description. There was a general appeal to the citizenry to contribute to the research resource of the company for the creation of biovalue. The value implied in the pervasive trust of Icelanders in the National Rescue Team is transferred to the activity of the company, symbolically conveying the message of the promises of rescue work that will be fulfilled in the future. By investing in biovalue, the citizens will be saved from the suffering of serious genetic diseases.

As Rose and Novas rightly point out, the kind of "biological citizenship" that they analyse operates within what they term "a 'political economy of hope'."[24] And the Icelandic case demonstrates clearly how this process is fuelled by strong expectations, belief in the medical fruits of population genetic research. And this belief partly explains the reaction to the critical voices that were raised against the national collection of biosamples. The strong expectations about medical progress that will eliminate the suffering that accompanies serious genetic illness should be beyond criticism. This has been well analyzed in a different context:

> There is an obvious continuity of exaggerated – sometimes wildly exaggerated – expectations. They do not depend upon or cling to single technologies or medical methodologies, but rather consume them as firewood in an ongoing, heated discourse. The discourse of hope is always ready to move on and leave behind the charred remains. There is no time for dwelling on the disappointments over unfulfilled miraculous scientific events. It is the

[20] Bellamy (2008: 2).

[21] Crick (1998: 9).

[22] Bellamy (2008: 2).

[23] Leydet (2006).

[24] Rose and Novas (2004: 453).

> nature, mechanics, and effects of this process that we have to understand. It is in this context that we have introduced the notion of "absolutist truth." An important effect of this new social concept of truth is the ability of hope to immunize itself against critique.[25]

The ability of the discourse of expectations to immunize itself against criticism is a key element in the economy of hope. And the authors continue their analysis which fits very well the Icelandic situation where the urgency of genetic research has been used to silence criticism of the methods that have been chosen to collect biomaterial for the research:

> the perceived urgency and necessity of finding new cures for harsh diseases make debates not only obsolete, but almost immoral, a worse-than-useless activity that consumes valuable energy, draws attention away from what is important, and potentially delays the timetable and progress of research. The result is the loss of a critical, normatively oriented political space. There is a call for immediate action, and not for words and deliberations.

The loss of a critically oriented political space is a serious matter in democratic societies and it is in relation to this point that the shortcomings of Rose's and Novas' analysis is most perspicuous. The vision of the citizen that emerges from their analysis of biological citizenship is twofold. On the one hand, the activist citizen who is engaged in furthering the cause related to his or her medical condition; and on the other hand, the citizens who are made up, for example, "as consumers of the potential range of goods which genomics has to offer".[26] The main logic driving them is the hope "for the production of wealth and health" and for the respective benefits to be reaped, either for individuals, companies or society. Against this backdrop, ethics serves as "a means of satisfying the values necessary to gain the trust and confidence of the citizen-consumer".[27] This picture of the citizen-consumer can hardly be associated with the vision of active citizenship in a democratic society which requires a different kind of participation.

Whether the process of formation and self-formation of the citizens in relation to biotechnology is characterized in terms of citizenship or not is, however, as such not a major issue. What is important is that this descriptive, allegedly "neutral" approach ignores the crucial question how democratic resistance to the formation of the citizen-consumer can be provided. How can the forces of social engineering fuelled by the economy of hope be effectively resisted? There is not much help in the biological citizenship analysis that help us to deal with that question which requires a much clearer distinction between consumer activity and democratic efforts than provided by Rose and Novas. In order to discuss the possibilities for democratic resistance to these processes we must consult a more normatively oriented theory which I will now turn to.

---

[25] Brekke and Sirnes (2011).

[26] Rose and Novas (2004: 457).

[27] Rose and Novas (2004: 458).

3.4 The Perspective of Deliberative Democratic Theory

The preceding analysis of biological citizenship is rooted in French post-structuralism with deep suspicion about projects to improve democracy. There is a wide theoretical gap between this Foucauldian approach and perspectives on scientific citizenship which draw upon deliberative democratic theory.[28] While forms of citizenship analyzed by Novas and Rose are mostly implicit, the attention is now paid to explicit attempts to involve the citizens in deliberative community engagement which presuppose active and reflective citizens in the spirit of Bernard Crick's project of citizenship education.[29]

In recent years, many kinds of deliberative events have been planned in order to engage citizens in relation to policy making in the area of biotechnology. The following is a typical description of an attempt to engage citizens in deliberation concerning a genome research project:

> Based on theories of deliberative democracy, our research group sought to engage community members in open dialogue. The approach assumes that once informed and provided with an opportunity for deliberation, non-experts members of the public are fully capable of reflecting on highly complex, technical issues. Participants work to understand each other and determine areas of agreement and disagreement; they both form and test their judgments within the collective discussion ... Our goal was not to correct a knowledge deficit, but to enable authentic deliberation and trade-offs among competing values. What risks will the public accept in order to gain the benefits of genomic research.[30]

This concise statement portrays well the intention of public engagement approaches in the spirit of deliberative democracy. As demonstrated by this example, the main emphasis in the application of deliberative democracy has been on "direct discussion and engagement".[31] The weaknesses of this approach are related to how difficult it is to successfully implement the idea.[32] But more importantly in this context, this kind of attempt to engage citizens in dialogical exercise is vulnerable to criticism in the spirit of the biological citizenship argument of Rose and Novas. The lure of this argument is that in a world of power play and conflict deliberative democratic ideas can appear naive.[33] A valuable reminder of the sceptical sociological view is that deliberative events can become diplomatic ways to ensure a more docile population through managed consultation. This is particularly dangerous when novel "promissory science", which citizens have no experience of but raises their hopes, is being introduced. This can lead to a premature justification of a policy with unforeseen social consequences. The evaluation of the benefits of such a project is very complex and it is difficult to frame these issues in a social debate.

---

[28] Abelson et al. (2003).

[29] Crick (1998).

[30] Hicks et al. (2008: 2).

[31] Irwin (2001: 16).

[32] Cf. Bickerstaff et al. (2010).

[33] Hagendijk and Irwin (2006: 169).

It is a cheap way of justifying a new genetic research project to ask in a deliberative forum: "What risks will the public accept in order to gain the benefits of genomic research".[34]

It is very misleading, however, to discuss deliberative practices mainly in terms of direct participation of citizens in various dialogical events and their impact on policy and decision making. Other crucial aspects of deliberative democracy are too often neglected in this discussion and need to be drawn upon in order to strengthen the vision of scientific citizenship. In their paper "Deliberating about bioethics", Gutmann and Thompson distinguish between the main social purposes served by deliberation in democracy. The first two of these purposes imply crucial aspects for democratic biopolitics and scientific citizenship: "to promote the legitimacy of collective decisions" and "to encourage public-spirited perspectives on public issues".[35] It is hoped that when citizens authentically engage in conversations about matters of common concern, they are likely to adopt a public standpoint, broaden their perspective, and be willing to revise their individual preferences in light of information and arguments. This makes real deliberation very different from all kinds of strategic interaction which people use in order to further their special interests. "When citizens bargain and negotiate, they may learn how better to get what they want. But when they deliberate, they can expand their knowledge, including their self-understanding as well as their collective understanding of what will best serve their fellow citizens".[36]

This deliberative view implies a vision of the citizen which is markedly different from the protection and the benefit views towards the citizens that have been dominant in bioethical discourse.[37] The main purpose of most ethical principles is to protect the personal domain of the individual from illegitimate intrusion and harm. This protection makes it possible to reap the benefits from advances in biotechnology and genetic research without interference from the citizens. In this way, these seemingly contrasting views complement each other and reinforce the passive role of citizens in relation scientific research. The protection view and the benefit view are important, but a public policy about biotechnology which is guided exclusively by these views provides no democratic resistance to the passive formation of the biological citizen as analyzed by Rose and Novas.

The deliberative view complements the protection and benefit views by taking into account the capacities or moral powers that are needed for active and reflective participation in a democratic society. If carefully used, such a guiding vision of scientific citizenship can resist the passive making of the biological citizen and even compensate for the loss of public value in that process. It is a duty of democratic politicians to conduct politics in such a way that the citizens are well informed and otherwise enabled to assume their responsibilities as democratic citizens which in turn should affect political decisions. Recall that according to Gutmann and

[34] Hicks et al. (2008: 2).

[35] Gutmann and Thompson (1997: 39).

[36] Gutmann and Thompson (1997: 40).

[37] Árnason (2009).

Thompson the first social purpose "served by deliberation in democracy" is promoting the democratic legitimacy of political decisions. Simone Chambers even sees accountability rather than consent as "the conceptual core of legitimacy".[38] This regards the way in which public policy is explained and justified to the citizens and thus relates to the activity of the legislators and policy makers. "Deliberative democratic theory critically investigates the quality, substance, and rationality of the arguments and reasons brought to defend policy and law".[39] This requires a scrutiny of working procedures and modes of reasoning in politics and public administration as well as of the ways in which information is presented and made accessible to the public, e.g. in the media. A crucial point here is that public accountability implies that policy can be justified to all that it affects and that they can accept its legitimacy.

These aspects of accountability can be regarded as necessary conditions for informed public opinion and deliberation about public policy and thus for scientific citizenship in a democratic society. The other main form that the promotion of democratic legitimacy of political decisions can take is to create deliberative forums designed to enhance and facilitate public deliberation about the issues for public policy. This aspect of participation as a feature of deliberative democracy should not, however, be overestimated and certainly not be confused with direct participatory democracy. In the latter, the primary emphasis is on widening the scope and increasing the amount of issues about which citizens can have a direct say. In deliberative democratic theory emphasized here, on the other hand, the main rationale behind the deliberative forum is to facilitate better informed and more reasonable decisions. The results of deliberative forums feed into policy decisions by informing policy makers about the arguments, viewpoints and concerns of the citizens.

3.5 The Icelandic Case Revisited

As described above, Rose and Novas use the Icelandic Health sector database (HSD) case to show how the Icelandic citizens were "a standing reserve for the creation of biovalue".[40] That is why they choose the Icelandic HSD case to exemplify biological citizenship. From the viewpoint of deliberative democratic theory, the HSD case is seen in a very different light and quite different questions about citizenship are raised in relation to that. That needs not imply that the skeptical sociological analysis is rejected, but it shows that it leaves out normative dimensions that are most important from a democratic perspective.

It has been argued that the parliamentary and public debates in Icelandic society about the Health Care Database do not amount to a democratic consent. Politicians failed in publicly articulating, explaining and justifying the policy regarding the database project. The public discussion was pervasive, but it was largely unin-

[38] Chambers (2003: 308).

[39] Chambers (2003: 309).

[40] Rose and Novas (2004: 455).

formed. There were close connections between the political sector and the company which conflated public and private interests.[41] All of this is crucial from the viewpoint of democratic citizenship. In the words of Chambers, from the deliberative viewpoint it is essential to investigate "the quality, substance, and rationality of the arguments and reasons brought to defend the policy and law" relating to the HSD project.[42] It is most important to improve governance by evaluating "the institutions, forums, venues, and public spaces available for deliberative justification and accountability".[43] Iceland is very weak in this regard. As mentioned above, there is no national ethical council in the country which has the task of raising public debate about bioethical developments and policy. Such institutions have the general aim of respecting the principles of scientific practice and moral conduct rather than serving the political authorities or the private sector. Deliberations in trustworthy bodies can both provide conditions for informed public debate and further democratic accountability.

Another indication of democratic weakness in Iceland is the media. It has been shown that with the exception of the population database project human genetic research and its potential effect on health and society has not been much discussed. The author is especially critical of the Icelandic media coverage which "has been framed in such a way that highly relevant questions are left unattended".[44] Another study demonstrates that news briefs from deCODE Genetics about their successes were uncritically taken up by the media which even leaves out the company's own reservations.[45] In this way, scientific and health care benefits of the research are treated as being beyond reasonable doubt, while uncertainty about financial issues is predominant.

Furthermore, it is especially significant, both from the perspective of deliberative democratic theory and the sceptical standpoint of biological citizenship, that the only community meetings held in order to discuss the population database project were organized by the genetic research company. These meetings were usually held in local health care centres and were conducted by the CEO of the company gathering public support for the database project.

All of this can be used to substantiate the statement of Rose and Novas about Icelanders as a standing reserve for the creation of biovalue. Icelandic society provides weak democratic resistance to the transforming powers of private companies, be they in the banking or the health sector. As a consequence, the Icelandic population is most vulnerable to the moulding forces of economy and biology and the vitality of the citizens is easily harnessed for the promises of creating health and wealth. According to the analysis of Rose and Novas, this makes the Icelanders a good example of biological citizenship. A fundamental flaw of this approach, however, is that it confuses descriptions of actual policies with rightful basis of action or public legitimacy which requires that "courses of action are chosen because they are

[41] Árnason and Árnason (2004).

[42] Chambers (2003: 309).

[43] Chambers (2003: 309).

[44] Hjörleifsson (2008: 75).

[45] Hjörleifsson et al. (2008).

based on reasons which all could accept".[46] This idea can be manipulated like any other but it provides an important vision of policy making in democratic society which the sceptical view of biological citizenship loses sight of. From the viewpoint of deliberative democratic theory, on the other hand, it shows both a lack of the major background conditions that enable people to exercise the capacities essential for scientific citizenship and the institutions necessary for public accountability.

Finally, I return to the example about the national campaign for the collection of biosamples described in the beginning of the paper. From the viewpoint of deliberative democracy on scientific citizenship, several questions need to be raised regarding public accountability and citizens' participation in this example. How was the campaign prepared, argued for and justified? What kind of community consultation was conducted before the collection? How well was the population informed about the project? What kind of public debate took place? And, more generally, how are the conditions in the public sphere for informed public opinion in Iceland? Dealing with these questions requires a scrutiny of working procedures and modes of reasoning in politics and administration as well as of the ways in which information is presented and made accessible to the public.

Such an analysis would assess the conditions from a democratic citizenship in Iceland that are radically different from the features of biological citizenship teased out from the sociological standpoint of Rose and Novas. In this way the two views on scientific citizenship discussed in this chapter are in tension, but that does not mean that these two analysis are in opposition. To the contrary, it is my contention that both perspectives are necessary to uncover the forces at play in contemporary biopolitics. A biopolitical analysis needs both a suspicious, sociological perspective which analyzes the processes of consumer-citizen formation fuelled by the "economy of hope" and a more normative and emancipatory viewpoint from which to criticize the ways in which policies are argued for, presented to, and debated by democratic citizens. While the former defies rational argumentation, the latter focuses on the conditions for democratic debate and accountability which is needed in a democratic public space.

3.6 Conclusion

I described an Icelandic case of a massive collection of biosamples and analyzed the example from two points of view. The former is a sociological theory of biological citizenship and the latter deliberative democratic theory which implies a very different notion of citizenship. While the former realistically demonstrates how citizens are implicated in the forces of social engineering, the latter provides the normative vision which facilitates democratic resistance to those forces. There is a fruitful tension between these two theoretical perspectives and they could have a complementary function in a fruitful biopolitical analysis.

[46] Held (2006: 245).

References

Abelson, J., P.-G. Forest, J. Eyles, P. Smith, E. Martin, and F.-P. Gauvin. 2003. Deliberations about deliberative methods: issues in the design and evaluation of public participation processes. *Social Science & Medicine* 57: 239–251.

Árnason, V. 2004. Coding and consent. Moral Challenges of the Database Project in Iceland. *Bioethics* 18: 39–61.

———. 2009. Scientific citizenship, benefit, and protection in population based research. In *Ethics of research biobanking*, ed. J.H. Solbakk, S. Holm, and B. Hoffman, 131–141. Berlin: Springer.

———. 2011. Databanks research: public and private interests. *Cambridge Quarterly of Healthcare Ethics* 20 (4): 563–571.

———. 2013. Scientific citizenship in a democratic society. *Public Understanding of Science* 22 (8): 927–940.

Árnason, V. (2016) Bioethics in Iceland. *Cambridge Quarterly of Healthcare Ethics* 25 (3): 421–434.

Árnason, V., and G. Árnason. 2004. Informed, democratic consent? The case of the Icelandic database. *Trames* 8 (1): 164–177.

Bellamy, R. 2008. *Citizenship. A very short introduction*. Oxford: Oxford University Press.

Bickerstaff, K., I. Lorenzoni, M. Jones, and N. Pidgeon. 2010. Locating scientific citizenship: The institutional context and cultures of public engagement. *Science, Technology & Human Values* 35 (4): 474–500.

Brekke, O.A., and T. Sirnes. 2011. Biosociality, biocitizenship and the new regime of hope and despair: Interpreting "portraits of hope" and the "Mehmet case". *New Genetics and Society* 30 (4): 347–374.

Chambers, S. 2003. Deliberative democratic theory. *Annual Review of Political Science* 6 (1): 307–326.

Crick, B. 1998. *Education for citizenship and the teaching of democracy in schools: Final report of the advisory group on citizenship*. London: Qualifications and Curriculum Authority.

Gutmann, A., and D. Thompson. 1997. Deliberating about bioethics. *The Hastings Center Report* 27 (3): 38–41.

Hagendijk, R., and A. Irwin. 2006. Public deliberation and governance: engaging with science and technology in contemporary Europe. *Minerva* 44 (2): 167–184.

Hakonarson, H., J.R. Gulcher, and K. Stefansson. 2003. deCODE Genetics Inc. company profile. *Pharmacogenomics* 4: 1–7.

Hedgecoe, A. 2004. *The politics of personalized medicine*. Cambridge: Cambridge University Press.

Held, D. 2006. *Models of democracy*. Cambridge: Polity Press.

Hicks, A., J.H. Hellyer, and B. Koenig. 2008. Involving the public in planning for the genomics revolution: An experiment in deliberative democracy. *The Bioethics Examiner* 11 (2): 1–3.

Hjörleifsson, S. 2008. Genetics, risk and medicalization. *A case study of preventive genetic technologies in Iceland*. Doctoral thesis, University of Bergen.

Hjörleifsson, S., V. Árnason, and E. Schei. 2008. Decoding the genetics debate – Hype and hope in Icelandic news media 2000 and 2004. *New Genetics and Society* 27 (4): 377–394.

Irwin, A. 2001. Constructing the scientific citizen: Science and democracy in the biosciences. *Public Understanding of Science* 10: 1–18.

Leydet, D. 2006. Citizenship. In *The Stanford encyclopedia of philosophy* (online). Available at: http://plato.stanford.edu/archives/win2006/entries/citizenship/

Rose, N., and C. Novas. 2004. Biological citizenship. In *Global assemblages: technology, politics, and ethics as anthropological problems*, ed. A. Ong and S.J. Collier, 439–463. Oxford: Blackwell Publishing.

Unesco Position Statement. 2004. *The plurality of literacy and its implications for policies and programs*. UNESCO Education Sector Position Paper, Paris.

Chapter 4
Chronic Disorders of Consciousness and *Homo Sacer*

Andrew Edgar

Abstract The purpose of this chapter is to develop a biopolitical response to certain ethical questions posed by chronic disorders of consciousness (CDoC). This response will draw on Giogio Agamben's account of *homo sacer* (Agamben G, Home Sacer: Sovereign Power and Bare Life (trans: Heller-Roazen D). Stanford University Press, Stanford, 1998). '*Homo sacer*' is a somewhat obscure term, found in Roman law, designating someone who has been expelled from the protection of the law. By addressing Agamben's own analysis of the 'overcoma' and the case of Karen Quinlan, it will be argued that a distinctive normative status can be ascribed to the CDoC patient: the CDoC patient exemplifies the condition of *homo sacer*, and as such of what Agamben calls 'bare life'. This argument poses a radical challenge to traditional approaches to bioethics and law. Precisely because *homo sacer* is placed outside the law, traditional legal or moral conceptions of rights are rendered inapplicable. However, it will also be argued that the Catholic tradition (defending the sanctity of life) or more utilitarian traditions (articulated in terms of quality of life) are rendered equally irrelevant. It will be concluded that the CDoC patient has moral status as a radical experience of bare life – and thus of what it is to be human – albeit one that cannot be expressed or articulated by the patient. The biopolitical challenge posed by CDoC thereby becomes that of facilitating the creative and poetic role of those who bear witness to the patient's experience. The biopolitics of CDoC thereby rests in the articulation for the patient of a radically different ethical status. The chapter will proceed by firstly offering an overview of the group of conditions classified as chronic disorders of consciousness, before outlining Agamben's biopolitics, his key terms, *zoē*, *bios* and 'bare life', and thereby explicating the idea of *homo sacer*. A series of Agamben's examples of *homo sacer* will be rehearsed, including that of the 'over-comatose' patient. This will provide the core material necessary for articulating a biopolitics of CDoC, and in particular to explicate the relationship between the patient and their 'witness'.

Keywords Agamben • Homo sacer • Bare life • CDoC • Illness experience

A. Edgar (✉)
Cardiff University, John Percival Building, Colum Drive, Cardiff CF10 3EU, UK
e-mail: edgar@cf.ac.uk

P. Kakuk (ed.), *Bioethics and Biopolitics*, Advancing Global Bioethics 8,
DOI 10.1007/978-3-319-66249-7_4

4.1 Chronic Disorders of Consciousness

The term 'chronic disorder of consciousness' embraces a number of conditions within which the patient's self-awareness is limited or absent. The permanent/persistent vegetative state (PVS) (Monti et al. 2010a), or unresponsive wakefulness syndrome (Laureys et al. 2010), is the most discussed example of a CDoC in the bioethics literature. This has been defined as 'wakefulness without awareness' (Jennett and Plum 1972) or more subtly as 'a complex neurological condition in which patients appear to be awake but show no sign of awareness of themselves or their environment' (Monti et al. 2010a). Despite a lack of responsive to the environment, the patient's body functions, so that they do not need artificial support such as a respirator, and the body will follow a cycle of sleeping and waking. The minimally conscious state (MCS) (Gosseries et. al. 2011), in which the patient is capable of brief periods of awareness, is also included in CDoC, as may conditions such as locked-in syndrome (where the patient is aware, but is unable to express that awareness) and chronic comas.

The moral problem of CDoC may be understood as lying in the challenge of articulating the nature of the condition. It appears to violate the usual categorical distinctions through which sense is ordinarily made of everyday experience. The patient in PVS is neither alive nor dead; the patient in MCS is neither conscious nor unconscious; neither an agent nor a passive object. The very identity of the person before and after the trauma that initiated the CDoC may be in doubt. The CDoC patient seemingly lacks memory of themselves prior to the trauma, and it is unclear how the intentions and preferences of that earlier person relate to or should be enacted on behalf of the patient.

The definition of condition may, further, be dependent upon the development of appropriate diagnostic techniques, be these behavioural measures (such as the Glasgow Coma Scale or more recent Coma Recovery Scale (Kalmar and Giacino 2005)) or more technically complex approaches such as brain imaging (including electroencephalography). Medical diagnosis may thus shift as different techniques are employed. But even if stable, diagnoses may not easily map on to the condition as it is experienced by relatives and lay observers of the patient (see Edgar et al. 2014). CDoC thereby becomes profoundly ambiguous, and the core concepts of traditional moral theory, such as 'autonomy', 'rights', 'dignity' and 'quality of life' cease to have an obvious or straightforward purchase.

4.2 Homo Sacer

Agamben finds the concept of '*homo sacer*' in the work of a second century Roman grammarian, Sextus Pompeius Festus. Here an archaic Roman law is recalled, according to which the judgement of a criminal to be '*home sacer*' (a sacred man) entails that: 'It is not permitted to sacrifice this man, yet he who kills him will not be condemned for homicide' (Agamben 1998, p. 71). *Homo sacer* may be killed, but not sacrificed (1998, p. 8). This is to assert that this human's life has no meaning, either within the judicial order (for their killing would not amount to murder), nor in terms of any other cultural dimension. The death of this human cannot acquire the meaning or status of a sacrifice, be it a sacrifice to the gods or a sacrifice for any other cause, such as, say, the preservation of the state. The human is thereby reduced to what Agamben calls 'bare life'.

Upon first encountering this example, it seems to be little more than a historical curiosity. Agamben does find similar legal conceptions in other jurisdictions, such as the Anglo-Saxon notion of the bandit or 'wolf's head' (*wulfesheud*). The bandit (epitomised by the legends of Robin Hood) stands outside the law, and significantly the very term 'wolf's head' is indicative of the expulsion of the human into an animal state (Agamben 1998, pp. 104–5). Yet Agamben's claim is not simply that the form of *homo sacer* may be found elsewhere. His argument is the more radical one that *homo sacer* is exemplary of the very structure of Western politics, and thus of biopolitics.

Agamben's account of biopolitics draws upon, but significantly develops, the work of Michel Foucault. Foucault presents biopower as a product of modernity. The exercise of sovereignty prior to the modern age is an exercise of power controlling 'legal subjects over whom the ultimate dominion was death' (1979, p. 142). Due to the restricted development of technology, with its limited control over natural resources and dangers, death confronted the pre-modern subject as a continual threat. The core power that the state has over the citizen is that of depriving them of life. In contrast, as the forces of production develop in modernity, biopower can be realised as the control and promotion of life (and not the mere imposition of death). As Foucault expresses this, biopower is exercised at the level of the body, and on life itself. The life of the individual body, or the body of the species, is brought into 'the realm of explicit calculations' (1979, p. 143). Exercised in order to promote economic productivity within capitalism, biopower adjusts the human subject to capital, inserting bodies into the machinery of production and adjusting populations to economic processes (1979, p. 141). Foucault summarizes the historical break so:

> For millennia, man remained what he was for Aristotle: a living animal with the additional capacity for a political existence; modern man is an animal whose politics places his existence as a living being in question. (1979, p. 143).

Agamben challenges Foucault's account by arguing that biopower is not unique to modernity. There is no historical rupture between pre-modern and modern power. Rather it is only in modernity that it becomes clear that biopower is fundamental to sovereignty. For Agamben: 'It can even be said that the production of a biopolitical body is the original activity of sovereign power' (1998 p. 6).

While Foucault has a unified and singular conception of 'life', as the subject of biopower, Agamben appeals to the ancient Greek distinction between *zoē* and *bios*. *Zoē* 'expressed the simple fact of living common to all living beings' (and hence the modern 'zoology'). In contrast, *bios* 'indicated the form or way of living proper to an individual or a group' (1998, p. 1). While on this initial account *bios* suggests a normative conception of the good life, more generally it is suggesting a life that is qualified or given meaning through participation in culture and politics. Hence, the most obvious English term grounded in *bios*, in this specific sense, is not 'biology' (which continues to appeal to common living), but rather 'biography'. A biographical life is one that has meaning and form, and thus is necessarily lived within a culture.[1]

The relationship of *zoē* and *bios* is further clarified, following Aristotle (in the *Politics* 1253a 10–18), as a distinction between 'voice' and 'language' (Agamben 1998, pp. 7–8). Animals have a voice, but this is merely the capacity to express pain and pleasure. It is only through language that the human being can articulate an understanding of the difference between justice and injustice. For Agamben, to have a language presupposes living within the *polis* (and thus within a human culture). The distinction between *zoē* and *bios* thereby allows Agamben to offer a subtle reinterpretation of Foucault's own appeal to Aristotle's definition of the human noted above: 'a living animal with the additional capacity for a political existence'. For Agamben, this 'additional capacity' is the 'exclusive inclusion' of *zoē* within the *polis* (1998, p. 7). It is precisely this paradoxical 'exclusive inclusion' that is fundamental to understanding Agamben's conception of biopolitics, and to his argument that biopolitics has, historically, underpinned all forms of Western state power.

By using to the formulation 'exclusive inclusion', Agamben is arguing that *zoē* is not simply included in politics, as the raw material that is shaped by biopolitical techniques. Language is not merely added on to the voice of the animal. Rather, the identity of the political body also rests upon a simultaneous exclusion of *zoē*. It marks a moment of violence that underpins the law. Here Agamben complements his account of Foucault through an appeal to the work of the Nazi jurist and political philosopher Carl Schmitt. Two elements of Schmitt's work are in play in Agamben's argument. Firstly, Schmitt argues that the distinction between friend and enemy lies at the centre of all political thought and action (2007). Schmitt's point is that a community acquires an identity and self-understanding of who it is only by recognising an enemy who it is not. The task of the state is that of ensuring the existence of an enemy. Crucially, the election of another as an enemy need have no grounding other that the will of the sovereign. The categorisation of someone as 'enemy' is an act of arbitrary violence.

The role of the other in securing one's identity parallels Foucault's accounts of madness and criminality. The rational and enlightened subject knows who and what they are precisely in their contrast to those who they are not. Thus, while retaining the form of Foucault's and Schmitt's argument, that identity is secured by focusing upon that which it is not, Agamben argues that Schmitt's friend/enemy distinction is not the fundamental one. It is, rather, the distinction between *zoē* and *bios* that is

[1] It may be noted that Aristotle's famous definition of the human being as *politikon zōon* does not undermine Agamben's interpretation. While humans may be social animals, they are not uniquely so. Ants, sparrows and lions are social. Yet such creatures lack a political life (*bios politikos*), which is to say a culturally chosen, developed and negotiated way of living together that aspires to be the good life (and not merely that which is determined by nature).

fundamental. *Bios* has identity precisely because it is not the form of *zoē* that is excluded from the political body. The insane and the criminal are thus potentially returned to *zoē* (or more precisely, what Agamben terms 'bare life').

Schmitt's second contribution to this analysis lies in his definition of the sovereign, as the one who decides upon the state of exception (Schmitt 2005; Agamben 1998, p. 11). This is to argue that sovereign power lies, not merely in the capacity to formulate law, but more fundamentally to suspend the rule of law, be this for society as a whole (as for example in the imposition of a state of emergency) or in the withdrawal of the protection of the law from an individual or group. For Foucault, the exercise of sovereignty is pre-modern, and thus prior to biopolitics. Agamben challenges this on two fronts. Firstly, he argues that liberal government, despite its self-understanding as being rationally grounded, and indeed as having given *zoē* its true and just form as *bios*, is underpinned by the violence of sovereignty. The rule of law is fragile, and can be suspended at any moment (see 1998, p. 9).[2] Secondly, Agamben reinterprets the Schmittean notion of sovereignty in biopolitical terms. Sovereignty is the capacity to strip the status of *bios* from the political subject, and thus to return them to a condition of *zoē*. Agamben's contention is that the declaration of the state of exception is an act of excluding *zoē* from the *polis*, and is thus fundamentally biopolitical.

It is here that the significance of *homo sacer* as expressive of the fundamental structure of biopolitics may be understood. *Home sacer* is the result of the exclusive inclusion. As such it has been abandoned by the state. The notion of 'abandonment' (and the 'ban') is crucial. *Homo sacer* is a subject who, prior to judgement, was included in the state, and thus was under the rule of law, as *bios* and as a possessor of language. Judgement upon them amounts to a decision, on the part of the sovereign, that the criminal is now an exception to usual legal processes. Their being excluded from the state's protection does not, however, entail simply that the rule of the law is removed from them. The very act of exclusion continues to include them within the law, for the category of '*homo sacer*' is itself a legal construct and an act of the sovereign. More precisely, *homo sacer* is not simply returned to a pre-political state of *zoē*. Rather, in being stripped of political status they are condemned to the (political) condition of 'bare life'. This term, 'bare life', is adopted from Walter Benjamin's essay 'The Critique of Violence' (1996), where it articulates the result of the sovereign violence that is seen to be fundamental to the exercise of the law (Agamben 1998, p. 65). In being judged, the language of *homo sacer* is no longer heard, and all biographical significance is stripped from them, so that neither their life nor death can have meaning.

[2] Agamben interprets the Hobbesian state of nature, not as a condition prior to the establishment of the state, but rather as 'a principle internal' to the state (1998, p. 105). It marks the dissolution of the state in the state of war of all against all, where 'the man is a wolf to men' (pp. 105–6). Yet it is also the fact that the sovereign is at once a role constituted by (and thus included in) the state and the social contract, and yet stands outside it (excluded) as the power to dissolve that contract.

4.3 Overcoma

In his exploration of the fundamental biopolitical structure of *homo sacer*, Agamben offers a series of examples or vignettes. The most relevant to the chronic disorder of consciousness is that of the 'overcoma' or *coma depassé* (1998, pp. 160f). What Agamben terms the classic coma 'is characterised by the loss of relational functions (consciousness, mobility, sensibility, reflexes)'. The overcoma, as defined by the French neurophysiologists Mollaret and Goulon, entails a 'total abolition of relational life functions correspond[ing] to an equally total abolition of vegetative life functions' (1998, p. 160). The continuing survival of the patient in an overcoma is therefore dependent upon the functioning of life support technology (maintaining respiration, and the functioning of vital organs). The overcoma patient is 'anatomy in motion', or even 'death in motion', for it amounts to a mere illustration of the physiological functioning of the body, no longer directed to the purpose of sustaining the patient's life (1998, p. 186).

Considered as a purely natural process, which is to say as the result of disease or injury, the coma patient has been turned back to zoē. They have been stripped of language, and if incapable of responding to stimuli from their environment, then they have lost voice as well. Yet Agamben's interest in the overcoma lies specifically in its construction as a political and cultural phenomenon. The 'overcoma' is a technological and legal category. The overcoma exists not because of natural processes, but because of advancements in medical technology that make this condition, that is neither life nor death, possible. Crucially this entails that the status of the patient, as either living subject or as corpse (and thus as to whether they posses to the right to further life-support), is a legal issue. Agamben seemingly delights in the defence offered by a man accused of murder. The victim does not die immediately, but falls into an overcoma after the assault. The victim dies only when doctors remove his life-support. The (unsuccessful) legal defence is that the assailant is innocent of murder, for it is the doctors who killed the victim (1998, p. 163).

The political position of the overcoma is further complicated by the development of organ transplant technology (and indeed in the above case, life-support is removed from the patient only after the heart is removed for transplantation). The overcoma patient is potentially an ideal organ donor, except for the fact that they are not, unproblematically, dead. This tension leads to the articulation of new definitions of death in the notion of 'brain death' (being the death of the entire brain) (1998, p. 162). Agamben finds this definition to be contradictory. While the diagnosis of brain death (which itself is dependent upon the development of appropriate diagnostic technology) comes to supersede the traditional criteria (the stopping of heartbeat and the cessation of breathing) that characterised systematic or somatic death, it is still defended on the grounds that 'most studies… demonstrated the inevitability of somatic death following brain death' (Agamben 1998, p. 163, citing Lamb 1988, p. 63). The concept of brain death is thus not free-standing. Brain death is death only because systematic death will occur shortly after the removal of life-support. Brain death exists only in terms of a pragmatic definition that facilitates the legal harvesting of transplantable organs. The biopolitics of technology and law thus combine to exclude the overcoma patient from the legal protection enjoyed by the living. The overcoma patient becomes not *zoē* (which is to say, the result of a purely natural process) but bare life (the product of politics).

It may be noted that the example of overcoma can be read as a further comment upon Foucault's historical distinction between sovereignty and biopolitics. For Foucault it is the sovereign whose ultimate sanction is that of killing their subject, while biopower is exercised to sustain life. In the case of the overcoma, biopower is indeed exercised to sustain the life of the organ recipient, but only by killing the donor. As *homo sacer*, the overcoma patient can be killed, but not murdered.

The case of Karen Quinlan further complicates this analysis. Quinlan was diagnosed as being in an overcoma when legal permission to remove life support was granted at the request of her parents. Here the court adjudged Quinlan to be already dead. After the removal of life-support, she continued to breath, and was thus recognised as being in a permanent vegetative state (PVS) (1998, pp. 163–4). Agamben does not explicitly refer to PVS, or indeed to CDoC. His interest in Quinlan's case is merely that it offers an extreme example of a body that has entered 'a zone of indetermination in which the words "life" and "death" have lost their meaning' (1998, p 164). What is significant here is that the CDoC patient cannot have life-support technology removed from them in order to facilitate organ transplantation. Their (systematic) death can only be brought about prematurely by denying them hydration and nutrition, and so allowing them to die, or by actively killing them. Their constitution as a biopolitical subject by medical technology (be this life-support or transplant technology) is thus of a different order to that of the overcoma patient. The involvement of only minimal technological intervention in sustaining their life serves to highlight the role of the law in the constitution of the CDoC patient.

To understand either the overcomatose or the CDoC patient as *homo sacer* is to argue that they can be killed but not sacrificed. As Agamben expresses this, if the patient is not alive (given the legal constructions suggested above) then they cannot be murdered (as the failure of the above murderer's defence demonstrated, for the doctors who removed life-support from the victim were not prosecuted for murder), and nor can they be sacrificed, for being already dead, there is no living being to sacrifice (1998, p. 165).

Orthodox bioethical approaches to the issue of CDoC may be articulated in terms of two broadly conceived strategies that challenge the implications of understanding the patient as *homo sacer*. On the one hand, it may be argued that the patient is unambiguously a living being – *bios* and not mere *zoē* – and as such either has the rights and legal protection of any other citizen, or that, as living, their life is sacred. Deontological and sanctity of life arguments may be marshalled in order to defend the patient, to conclude that the killing of the patient is, indeed, murder. On the other hand, it may be argued that the patient can be sacrificed. A quality of life argument can be formulated to the effect that the patient's body and life are no longer of any use to them (for lacking consciousness, they can have no quality of life). Their body (or at least its organs) can be of use to others, and will contribute to raising the organ recipient's quality of life. According to this argument the patient can be killed, but crucially their death can be given meaning, for their life is being sacrificed to the cause of another's medical treatment. Agamben himself does not explore these arguments beyond noting the tension between positions defending the patient as alive and as dead (1998, p. 164). However, his analysis of *homo sacer* begins to highlight the fundamental inadequacy of these traditional bioethical approaches.

The antimony of deontology and consequentialist ethics, whereby seemingly valid arguments can be given both in favour of and against supporting the life of the patient, can be understood as a result of biopolitics. This is to argue that bioethics reproduces the structure of sovereign violence. It was noted above that Agamben criticises liberalism for assuming that it has secured the just *bios* for *zoē*. The liberal state is the normatively appropriate form of human life. Thus, liberal conceptions such as human rights and the rule of law are assumed to be rationally defensible as normative absolutes. For Agamben this assumption denies the sovereign violence (expressed, by Schmitt, both in the capacity to suspend the rule of law, and in the arbitrary designation of an enemy). Crucially law, as a political structure underpinned by sovereignty, is understood by Agamben, following Schmitt, as a purely positive structure, which is to say a structure without normative or moral content. Legal categories do not identify an entity that exists prior to and outside the law, and nor does the legal judgement express a moral truth. Rather they constitute, in an act of accusation, an entity within the law (see Agamben 2008). In a bioethics that is homologous with legal thinking, such as traditional deontology and consequentialism, antimonies occur because each side of the antimony is grounded in a different and opposite, but equally arbitrary, categorisation of the legal subject (in this case, the CDoC patient). The failure of traditional ethics to respond coherently to CDoC is thus expressive of the biopolitical nature of legal and ethical thinking.

While this analysis may identify a problem in the bioethics of CDoC, it does little to resolve it. In practice, Agamben's own ethics remain tentative and critical, for while they are articulated in terms of the Aristotelian question of the good life, they appeal to a 'coming community', and a Messianic suspension of the law (and not, for example, to the ethics of contemporary or actual communities as might be found in communitarian approaches) (see Agamben 1993a). Yet the text of *Homo Sacer* does nonetheless offer further resources for understanding CDoC, and for articulating the ethical challenges that the conditions pose.

4.4 The Muselmann

Two further vignettes of *homo sacer* will be briefly reviewed as having relevance to understanding CDoC. These are a detail of ancient Roman law, *vitae necisque potestas*, where a father had power of life or death over a son; and the *Muselmann*, the most extreme victim of the death camp.

Under the law of *vitae necisque potestas* the Roman citizen had an absolute power of life and death over his sons, and absolute in the sense that this is neither the responsibility of the father to punish the son's criminality, nor yet a product of the structure of the patriarchal family (Agamben 1998, p. 87). Rather, the Roman citizen, as a son, enters political life as someone who can be killed. That is to say, they enter politics not as natural life (*zoē*) but rather as bare life, someone who has at once been defined legally, but thereby excluded from legal protection. As such, every citizen is at least initially *homo sacer* (1998, p. 88).

Such an example may initially seem to have little to do with CDoC patients. Yet, as the example of Quinlan indicated, the survival of the patient may be dependent upon the decision of their relatives and carers. This may be articulated in terms of the deep and tragic problem of how relatives and carers can come to understand, with any assurance, what is the just way in which to treat the patient. The focus of the problem lies in the fact that the CDoC patient has typically lost language. While a patient in a state of minimal consciousness may have moments of lucidity, and thus language use, this will be rare or absent in the case of PVS patients. At best, the patient will have a voice, in the sense used above, in so far as they can express pain and pleasure in response to stimuli. The interpretation of this voice, and its translation into language, becomes a challenge to which relatives and carers respond (see Edgar et al. 2014). This interpretation may be expressed in attempts to discover a language for the patient by finding in the patient's movements and gestures an expression of, not merely their pain or pleasure, but rather their sense of justice concerning the condition they are in. Here the relative and carer must struggle with metaphysical issues of personal identity, as the intentions and preferences of the person prior to the trauma that delivered them into a CDoC are reconstructed and mapped, often awkwardly and with difficulty, onto the voice, the movements and gestures, of the patient.[3] The relative is left with questions as to whether the patient, as they are now, would want earlier intentions (expressed perhaps in a living will) to be put into action, and thus whether the patient now has any significant continuity with the pre-traumatic person. Also the patient acquires a proxy language through the relatives' and carers' continuing interaction with them, by talking to them, responding to and observing them, taking them out into the public world beyond the hospital ward, and thus by presenting them, through the care taken in arranging their clothing, hair or make-up, as prepared for social interaction.

The carers and relatives thus construct and maintain what may guardedly be described as an 'illusion' of a *bios* for the patient.[4] They struggle to sustain meaning for the patient, as a continuing member of a family and wider community, in the

[3] Technology complicates this issue somewhat. Research using brain imaging techniques has suggested that some PVS patients may have an active mental life that they cannot express (Monti et. al. 2010a, b). This can involve responding to requests to think of specific events (such as to imagine playing tennis). Such research both compounds the biopolitical construction of the CDoC patient as a technological subject, and give further, often perplexing and ambiguous material by which the patient's relatives can expected to interpret the patient's intentions and attitudes, which is to say, to construct a language for them.

[4] 'Illusion' here is used in the Hegelian sense of the German term '*Schein*' (see Hegel 1973). As such, it is a distorted representation of the truth (as opposed to a delusion, which is a deception, offering a representation of something that does not exist).

face of bare life. Yet crucially the patient ultimately has no language in order to participate in decisions about their care, or indeed their continued existence. The patient is unable to protest the justice or injustice of their own condition, to deny or affirm the interpretations that are imposed upon them. Thus the decision as to their continued existence falls upon those who can use language, the relatives and carers. If the carers' understanding of the patient's voice is inadequate, or is a projection of their own quite understandable desires and frustrations, then the *bios* that the carer attributes to the patient may approach the same arbitrariness that was seen in the sovereign decision on the state of exception. The CDoC patient becomes *homo sacer* to their relatives, akin to the Roman son to his father. The example of the *Muselmann* will help to clarify further the nature and implications of this observation.

The *Muselmann* (where the word is a derogatory term for 'Muslim') is for Agamben the most extreme figure within the Nazi death camps. (Agamben draws significantly on the writings of Primo Levi in articulating the nature of the *Muselmann*.) The death camp exemplifies much of the structure of modern biopolitics (and Agamben criticises Foucault for making the hospital and asylum, and not the camp, the institutional focuses of his account of biopower (Agamben 1998, p. 4)). Under Nazism the Jew has the protection of the law systematically removed from them, rendering them *homo sacer*. Yet within the camp, the *Muselmann* experiences an even more extreme subjugation. The *Muselmann* loses the will to survive. Their humiliation and fear strips from them consciousness and personality, rendering them utterly apathetic. They are excluded even from the life of the camp, unable to communicate and shunned by other inhabitants. 'Mute and absolutely alone, he has passed into another world without memory and without grief' (Agamben 1998, p. 185). The *Muselmann* lacks language, but in their apathy even their voice is threatened. They are unable to distinguish 'between the pangs of cold and the ferocity of the SS' (ibid.).

While the CDoC patient has not undergone the systematic and intentional acts of degradation suffered by the *Muselmann*, for, as noted above, the CDoC patient is typically the victim of disease or injury and their status as bare life (as opposed to natural *zoē*) rests upon the legal construction of that condition. Yet there is still something in common. Both CDoC patient and the *Muselmann* experience bare life at its extreme. Both are stripped of language. Conceivably the CDoC patient is no more aware of the difference between human and natural stimuli than is the *Muselmann*. It is this extreme experience that is of normative importance for Agamben, and through which something akin to an ethical imperative can be derived from his analysis, and this in the role of the 'witness'.

The figure of the *Muselmann* is complemented by that of the witness. Agamben notes that a number of survivors of the camps are motivated by a desire to bear witness to what they have experienced (and indeed, it is exactly this witness that the Holocaust denier seeks to invalidate) (1999, p. 15). Yet, precisely by surviving the camp, the witness has not had the extreme experience of the *Muselmann*. They have not descended into those depths of inhumanity. The *Muselmann*'s experience is the most profound and penetrating. It is, as it were, the pure experience of bare life. But the *Muselmann*, robbed of language, cannot explicate that experience. Muselmann and witness thus form a two-part structure, who 'like the tutor and the incapable person and the creator and his material, are inseparable; their unity-difference alone constitutes testimony' (1999, p. 150). The survivor bearing witness not to the camp itself but to the experience of the *Muselmann* , and that an experience that they did not share and that cannot, strictly, be spoken, 'speaks only on the basis of an impossibility of speaking' (1999, p. 164).

The moral status of the CDoC patient may be understood in these terms. The patient, like the *Muselmann*, has the most extreme experience of bare life. Agamben may be interpreted here as offering something akin to a sanctity of life argument. Yet, the moral status of the patient lies not in the fact that they are alive (for they are only ambiguously so) and nor yet in that their existence is sanctified. They are not 'sacred' in the usual meaning of the term, but only in the obscure and ambiguous sense of *sacer* (see Agamben 1998, p. 78–9 on the interpretation of *sacer*). The moral status of the patient lies, rather, in the fact that they experience bare life and thus what it is to be human in the face of biopower. Like the *Muselmann*, it may be said of the CDoC patient that: 'the human being is the inhuman; the one whose humanity is completely destroyed is the one who is truly human' (1999, p. 133). As such, the CDoC patient requires a witness.

Agamben is arguing that experience cannot be adequately expressed in language. There is always a remnant. The failing of law, and a legally based bioethics, is that it does not acknowledge the fact that legal categories do not grasp the non-legal reality – legal thinking does not recognise a remnant. It does not recognise the constitutive moment of accusation, and as such assumes that the particular case, the experience of the unique and individual patient, may be adequately incorporated into the broad categories of law. An ethics of CDoC must, in contrast, be an ethics of witness. It would lie in the frequently frustrating and perhaps even futile struggle to find a language, unique and appropriate, to the individual patient, and one that does away with the traditional, legalistic, categories of bioethical thought, such as dignity and respect (see Agamben 1999, p. 69). It will offer a poetry that is shocking. It is shocking precisely in that it will expose the fact that it has ultimately failed to grasp the experience of the CDoC patient. It will be poetry of 'an unprecedented lack of experience', as Agamben describes the poetry of Baudelaire (1993b, p. 47). As such, the struggle of the relative or carer to find a language to articulate the patient's experience, and so to sustain a *bios* around the patient, will be one in which the relative and carer are aware of their own desires, biases and frustrations, and thus of how the patient continually escapes them. While the *bios* that is constructed may be an illusion, it is a necessary illusion, and the only thing upon which a genuinely ethical debate, and as such an aspiration to an understanding of CDoC outside the abandonment of contemporary legal structures and biopolitics, can be performed. A shocking poetry of witness will continue to challenge the illusory meanings constructed, maintaining a dynamism and invention in ethical debate.

References

Agamben, Giorgio. 1993a. *The Coming Community* . Trans. H. Michael. Minneapolis: University of Minnesota Press.

———. 1993b. *Infancy and History: The Destruction of Experience.* Trans. L Heron. London: Verso.

———. 1998. *Home Sacer: Sovereign Power and Bare Life.* Trans. D. Heller-Roazen. Stanford, CA: Stanford University Press.

———. 1999. *Remnants of Auschwitz: The Witness and the Archive.* Trans. D. Heller-Roazen. New York: Zone.

———. 2008. K. In *The work of giorgio agamben: Law, literature,life*, ed. Justin Clemens, Nicholas Heron, and Alex Murray, 13–17. Edinburgh, Edinburgh University Press.

Benjamin, Walter. 1996. 'Critique of Violence. In *Selected writings, volume one 1913–1926*, ed. Marcus Bullock and Michael W. Jennings, 236–252. Cambridge, MA: Belknap Press.

Edgar, A., C. Kitzinger, and J. Kitzinger. 2014. Interpreting chronic disorders of consciousness: medical science and family experience. *Journal of Evaluation in Clinical Practice* 21: 374–379.

Foucault, Michel. 1979. *The history of sexuality volume 1: An introduction.* London: Allen Lane.

Gosseries, O., A. Vanhaudenhuyse, M. Bruno, A. Demertzi, C. Schnakers, M.M. Boly, A. Maudoux, G. Moonen, and S. Laureys. 2011. Disorders of consciousness: coma, vegetative and minimally conscious states. In *States of Consciousness: Experimental Insights into Meditation, Waking, Sleep and Dreams*, ed. D. Cvetkovic and I. Cosic, 29–55. New York: Springer.

Hegel, G.W.F. 1973. *Hegel's Aesthetics.* Trans. T.M. Knox. Oxford: Clarenden Press.

Jennett, B., and F. Plum. 1972. Persistent vegetative state after brain damage: A syndrome in search of a name. *The Lancet* 299: 734–737.

Kalmar, K., and J.T. Giacino. 2005. The JFK coma recovery scale – revised. *Neuropsychological Rehabilitation* 15 (3/4): 454–460.

Lamb, D. 1988. *Death, brain death and ethics.* London: Routledge.

Laureys, S., G. Celesia, F. Cohadon, et al. 2010. Unresponsive wakefulness syndrome: a new name for the vegetative state or apallic syndrome and the European task force on disorders of consciousness. *BMC Medicine* 8: 68.

Monti, M.M., S. Laureys, and A.M. Owen. 2010a. The vegetative state. *British Medical Journal* 341: 292–296.

Monti, M.M., A. Vanhaudenhuyse, M.R. Coleman, M. Boly, J.D. Pickard, L. Tshibanda, A.M. Owen, and S. Laureys. 2010b. Willful modulation of brain activity in disorders of consciousness. *The New England Journal of Medicine* 362 (7): 579–589.

Schmitt, C. 2005. *Political Theology. Four Chapters on the Concept of Sovereignty.* Trans. G. Schwab. Chicago: University of Chicago Press.

———. 2007. *The Concept of the Political.* Trans. G. Schwab. Chicago: University of Chicago Press.

Chapter 5
Biopolitics and the Longevity of Left-Handers

Gardar Arnason

Abstract One aspect of Foucault's concept of biopolitics concerns how society manages or governs certain groups of people, in particular when that sort of governance requires scientific knowledge about the group or kind of people in question. Left-handers are one example of a kind of people that has to be studied and managed. In the early twentieth century, left-handedness among school children was perceived as a growing problem. Not only parents, but teachers and school principals had to deal with this problem and they called upon scientists to give them more knowledge about these problematic children. Since then there has been no end to research on left-handers, in particular in psychology and health sciences. The paper discusses one such case, a series of studies aimed at answering the question whether left-handers die on average younger than right-handers. Discussion of these studies reveals how problematic the scientific definition of left-handedness is and also what effects (what Foucault termed power-effects) this sort of research has on its subjects. Although there is nowadays less pressure in the school system to manage left-handed children, the production of knowledge about left-handers continues. The paper focuses on how this knowledge is diffused through scientific literature, popular science literature and mass media; and how these studies have given rise to confrontation and conflict.

Keywords Foucault • Biopower • Life-expectancy • Abnormality • Classification

5.1 Introduction

In the first volume of *The History of Sexuality*, Michel Foucault introduces the term "biopower" to describe a new kind of power relations which began to emerge in the seventeenth century.[1] Biopower is power over life; it has to do with how individuals

[1] Foucault (1990, pp. 139–140).

G. Arnason (✉)
Institute for Ethics and History of Medicine, University of Tübingen,
Gartenstrasse 47, 72074 Tübingen, Germany
e-mail: gardar.arnason@uni-tuebingen.de

P. Kakuk (ed.), *Bioethics and Biopolitics*, Advancing Global Bioethics 8,
DOI 10.1007/978-3-319-66249-7_5

and groups are governed in modern societies. According to Foucault, biopower resulted from the merger of two forms of power, or two poles, which he calls "anatomo-politics of the human body" and "biopolitics of the population". The anatomo-politics of the human body "centered on the body as a machine: its disciplining, the optimization of its capabilities, the extortion of its forces, the parallel increase of its usefulness and its docility, its integration into systems of efficient and economic controls."[2] The second pole within biopower, biopolitics of the population, "focused on the species body, the body imbued with the mechanics of life and serving as the basis of the biological processes: propagation, births and mortality, the level of health, life expectancy and longevity, with all the conditions that can cause these to vary."[3]

The sort of power, which Foucault has in mind, is not physical force, it is not the top-down power of the sovereign over the subjects, it is not violence. Power, in Foucault's sense, is more like a network, where various forces both constrain what one can do, think, and be, and also create the possibility of acting in a certain manner, thinking certain type of thoughts and being in certain ways. Power is thus not merely limiting or constraining, it is creative and driving. Modern power could be describes as a network of people, institutions and practices, which creates and structures the space of possible action, thought, and being.[4] In order to govern people, it becomes necessary to know them. Knowledge and power go hand in hand; power requires knowledge and knowledge has power effects. Knowledge about certain groups or kinds of people is particularly in demand. As a society, we need to understand criminals and their different kinds in order to know what to do with them, same with such kinds of people as children, refugees, teenage mothers, homosexuals and, of course, those who are defined by their illness or disability, such as various types of addicts, autistic people, the depressed, the obese, sociopaths and psychopaths, and the list goes on and on.[5]

The two poles within biopower, which we can term "disciplinary power" and "biopolitics", both drive knowledge production about various kinds of people to help manage them. Left-handers are one example of a kind of people that has to be studied and managed. In the early twentieth century, left-handedness among school children was perceived as a growing problem. Not only parents, but teachers and school principals had to deal with this problem and they called upon scientists to give them more knowledge about these problematic children. Since then there has been no end to research on left-handers, in particular in psychology and health sciences. I will discuss one such case, a series of studies aimed at answering the question whether left-handers die on average younger than right-handers. My discussion of these studies reveals how problematic the scientific definition of left-

---

[2] *Ibid*, p. 139.

[3] Ibid, p. 139.

[4] See Foucault (1980), in particular "Two Lectures", pp. 78–108.

[5] Ian Hacking has further developed the idea of kinds of people as objects of knowledge and power, see in particular "Making Up People" reprinted in Ian Hacking (2002, pp. 99–114; Ian Hacking 1999), especially Chaps. 4 and 5; and Ian Hacking (2007, pp. 285–318).

handedness is and also what effects (what Foucault termed power-effects) this sort of research can have on its subjects. Although there is nowadays less pressure in the school system to manage left-handed children, the production of knowledge about left-handers continues. I look at how this knowledge is diffused through scientific literature, popular science literature and mass media; and how these studies have given rise to confrontation and conflict.

5.2 Pre-scientific Conceptions of Left-Handedness

There were left-handers long before there was science to study them. Left-handers must always have stuck out; they would use their left hand to eat and to hold things, whereas the vast majority would prefer to use their right hand. This preference will have become very noticeable with the use of tools and problematic with the emergence of complex tools, where there would be a "correct way" of holding the tool. Although speculations about the emergence of handedness and its association with *right* and *wrong* tend to focus on the use of tools,[6] a different anthropological observation may go some way to explain the association of handedness with *good* and *bad*. In many cultures the left hand is used to clean oneself after defecating, with the result that the left hand is not allowed to touch food when eating, for sensible hygienic reasons. As the left hand becomes the dirty hand, the right hand is the clean hand, and only the right hand can be used for eating and for saluting others.[7] The association of right and left with good and bad appears immediately.

Long before the emergence of the normalising society in the nineteenth century, before left-handedness became bad as an *abnormality*, left-handedness was considered wrong and bad, apparently across history and cultures. The universality of negative associations with left-handedness is quite astounding. Aristotle, for example, says in his *Metaphysics* (986a15), that the Pythagoreans "hold that there are ten principles, which they enunciate in a series of corresponding pairs: (1) Limit and the Unlimited; (2) Odd and Even; (3) Unity and Plurality; (4) Right and Left; (5) Male and Female; (6) Rest and Motion; (7) Straight and Crooked; (8) Light and Darkness; (9) Good and Evil; (10) Square and Oblong." Here, left is associated with other terms considered negative, such as "dark", "evil" and "female".[8] Anthropologists have found similar negative associations with the term "left" in various cultures, such as the Nyoro of East Africa and the Gogo of Tanzania[9]; the Nuer of South Sudan and the Temne of Sierra Leone[10]; the Maori of New Zealand and the

[6] See for example Corballis (1991, p. 104) and Uomini (2009).

[7] See Binnie Dawson (1977, p. 426).

[8] The citation is from G. Cyril Armstrong's translation of the *Metaphysics*, available at the Perseus Catalogue website: http://catalog.perseus.org/catalog/urn:cts:greekLit:tlg0086.tlg025.perseus-eng1

[9] Corballis (1983, pp. 1–2).

[10] Binnie Dawson (1977, p. 425).

Wulwanga of Australia, to name just a few.[11] Robert Hertz summarized the view of the left hand in different cultures thus: "To the right hand go honours, flattering designations, prerogatives: it acts, orders, and takes. The left hand, on the contrary, is despised and reduced to the role of a humble auxiliary: by itself it can do nothing; it helps, it supports, it holds."[12]

According to a common Greek lexicon the Greek word for "left," *aristeros*, can mean "ill-boding" or "ominous" and the word for "right," *dexios*, can mean "boding good" or "fortunate." In way of explanation, the lexicon claims that this "sense came from the Greek augurs looking to the North, so that lucky omens, which came from the East, were on the right, while the unlucky ones from the West were on the left."[13] In 270 B.C. the Greek poet Theocritus wrote "Lo there! a twitch o' my right eye. Shall I be seeing her?" and almost 2000 years later John Gay wrote in one of his fables: "That raven on your left-hand oak/(Curse his ill-betiding croak)/Bodes me no good."[14] From Latin we have the word sinister, and left-handers are sometimes referred to in the scientific literature as "sinistrals". The Latin word for right is *dexter*, and in English we call those who are skillful with their hands "dexterous". In French the word for left is *gauche*, which we also use in English for what is awkward or inappropriate. Then there are left-handed compliments. In German *links liegen lassen* (leave to the left side) means to ignore someone. In 1946, psychiatrist Abram Blau noted "in language, negativism is quite recognizable in the connotations of contrariness associated with the various words for the left side, such as gauche, sinister, left-handed compliment, left (radical) party, etc."[15] It may seem odd to include the "left (radical) party", since the political term "left" is a historical coincidence: The French National Assembly decided in 1789 to sit radicals on the president's left-hand side, the conservative nobles on his right and the moderates in the middle. This historical coincidence is just the bad luck of the political left. Psychologist Chris McManus has suggested, like Blau, that "left" in politics is a negative term, on the grounds that it is one member of the pair right-left, where right is symbolically the positive term and the norm, while left is the negative term and the deviation. McManus then argues that since people, when asked about their views on anything, are more likely to answer with the positive term than the negative term, when an answer requires one or the other (and everything else being equal), people are more likely to claim to be on the right wing in politics than the left. Hence the left will always be in the minority and the right in the majority.[16] If

[11] Hertz (2013, pp. 343–345). The essay was first published in 1909 as "La prééminence de la main droite: étude sur la polarité religieuse," *Revue Philosophique* 68, pp. 553–580.

[12] Hertz (2013, p. 335).

[13] Liddell and Scott (1992).

[14] Coren (1992, p. 11). The quote from Theocritus can be found in Cholmeley (1919), 3rd Idyllium. The quote from John Gay can be found in Gay (2005, p. 97) (Fable XXXVII: The Farmer's Wife and the Raven).

[15] Blau (1946, p. 91).

[16] McManus (2002, pp. 260–264).

McManus is right, the negative connotations of the word "left" have direct party-political consequences.

Left-handedness has had its place in the dualism manifested in the endless double-column of good and bad, but through the explosion of statistics and statistical reasoning in the early nineteenth century, left-handedness became bad in a new way: as a deviation from a norm, as an abnormality. The statistical language of norms and deviations does not, at the mathematical level, entail any value judgments, but it evolved as a way of organizing knowledge about human affairs and quickly the statistical norm became also an ideal. At the same time abnormalities and deviations become something to correct, cure or prevent. In a normalising society, left-handedness is an obvious target for prevention and correction, but for that, we need knowledge about left-handers.

5.3 The Scientific Study of Left-Handers

Until recently, children in the Western world were chastised if they held tools such as a pen or a fork in their left hand. At school, left-handed children were forced to hold the pen in their right hand. It was simply a practical matter of school room discipline. At the beginning of the twentieth century, something changed. Educators started to ask for knowledge about left-handers, in order to be better able to deal with them. Some thought left-handedness was on the rise among pupils and they were worried. In 1914 the *Teacher* published an article titled "What shall we do with left-handed pupils?" In the article A. B. Poland, school superintendent in Newark, New Jersey, is quoted saying: "I have not adequate knowledge derived from experience to warrant me in expressing an opinion as to the best means of dealing with left-handed pupils."[17]

Left-handers or, in this case, at least left-handed pupils, were a group that needed to be dealt with, they had to be controlled, managed, disciplined. In order to control, manage and discipline them more effectively, it became important, as the superintendent noted, to gain some knowledge of this sort of people. Abram Blau, who I mentioned in the previous section, was a psychiatrist at the New York University College of Medicine and before that a school psychiatrist at the Bureau of Child Guidance, Board of Education in New York City. Blau wrote a book about handedness, which was published in 1946 with the title *The Master Hand*, where he notes that left-handedness is an increasing problem:

> A principal of an elementary public school in one of the better neighbourhoods in New York City remarked to me about the alarming increase of left-handedness among his pupils. This was undoubtedly due to the increased cultural tolerance of sinistral tendencies which should really be discouraged and the child trained to adjust himself normally to this right-handed world.[18]

[17] McMullin (1914); quoted in Harris (1980, p. 3).

[18] Blau (1946, p. 90).

The psychological causes of these "sinistral tendencies" where quite clear to Abram Blau, namely the child's contrary attitudes:

> Sinistrality is the product of a contrary attitude on the part of the infant and young child. In other words, sinistrality is thus a symptom or manifestation of an attitude of opposition or negativism along with such other signs as disobedience, refusal to eat, temper tantrums, rebelliousness, etc. In place of a wish to comply with the social and cultural pressures toward the use of the right hand, there exists an active attitude of opposition which manifests itself in the development of sinistrality.[19]

This view was not unusual, neither in academic writing nor in society in general. A few years before, the British educational psychologist Sir Cyril Burt had written in his book *The Backward Child* that left-handers were willful: "Even left-handed girls [...] often possess a strong, self-willed and almost masculine disposition."[20] Burt took left-handedness quite literally as the opposite of dexterity, describing left-handers thus: "They squint, they stammer, they shuffle and shamble, they flounder about like seals out of water. Awkward in the house, and clumsy in their games, they are fumblers and bunglers at whatever they do."[21]

From the perspective of biopolitics, one can identify the need to manage a certain group of people, in this case left-handed pupils, and how that gives rise to demand for scientific knowledge about that group. The scientific knowledge in turn presents left-handers as problematic, they are abnormal, a deviation, and this reinforces the need to deal with them.

Nowadays the incitement for knowledge production about left-handers is more economic than political. The political need to deal with left-handers has changed in Western schools; there is no explicit demand for knowledge about left-handers for the purpose of converting them to right-handedness, but there is still a need to manage them, as is borne out by recent books bearing titles such as *Your Left-Handed Child: Making Things Easy for Left-Handers in a Right-Handed World* and *Loving Lefties: How to Raise Your Left-Handed Child in a Right-Handed World*, not to mention books teaching left-handed pupils to write properly, such as *Cursive Writing Skills for Left-Handed Students*, *Left Hand Writing Skills* (three books), and *Handwriting Program for Cursive Left Hand (Preventing Academic Failure)*. I suspect that this change of educational policy is not only stemming from increased tolerance and a more positive view of left-handers, but even more from the fact that left-handedness is not easily "rectified" without applying certain physical force and even physical constraint over an extended period of time—and such methods are out of fashion. Although educators are nowadays not willing, or even able within the law, to constrain pupils' left hand in order to make them right-handed, there are still people who promote making left-handed children "normal", without the application of physical force or constraint. Just to give one example, a popular German advice book for parents of young children suggests that left-handed children should be "given the opportunity to learn to write with the right hand", not through coercion

[19] Blau (1946, p. 91).

[20] Burt (1937). Quoted in Corballis (1991, p. 91).

[21] *Ibid.*

but "with patience and care" so that the child "is proud to be able to write 'like the others'".[22]

The school masters and superintendents may not be demanding knowledge about left-handers in order to convert them to right-handers, but there is still a need to deal with left-handers, particularly left-handed pupils but also left-handed employees, athletes, musicians, cooks, gardeners and left-handed consumers in general, and this need to deal with them also creates a need for knowledge about left-handedness. That there is an economic incitement for knowledge about left-handers is manifested by the great number of books aimed at left-handers and at parents of left-handed children, as well as the proliferation of stores, both virtual and physical, selling products for left-handers. The online retailer Amazon lists over 700 books with the word "left-handed" in the title. Not only left-handers themselves are interested in knowledge about left-handers; parents, teachers, physicians, employers, designers, manufacturers, and service providers have to deal with left-handedness and left-handers, and this creates a need for knowledge.

Although the need for knowledge about left-handers has changed, the scientific research on left-handedness has probably never been more productive than during the last three decades. It is difficult to explain this scientific interest in left-handedness without referring to the need to "deal with" left-handedness, and the political and economic incitement to which knowledge about left-handers is subjected. Another motivation for left-handers research is the historical baggage of the concept: left-handedness is interesting because it is deviant, abnormal and, after all, perhaps not quite *right*.

Many scientists studying left-handedness acknowledge that handedness is just a part of a larger phenomenon, namely the laterality of the body, but they have nonetheless a very strong tendency to talk about their studies and results as if the subject is strictly left-handers, or they even explicitly limit their study on human laterality to left-handedness. Most of the research on laterality is concerned with deviations from the usual right shift in human laterality (the tendency of humans to be right-sided). When classifying the subjects, questionnaires are used which have mostly, or even exclusively, items regarding hand preference. Much more research is done on left-handedness in particular than the common overall right-sidedness or the lateral asymmetry of the brain and the body. Much of this research aims at establishing statistical links between left-handedness and anything human, despite the fact that not much is known about the causes of left-handedness, nor right-handedness for that matter, and thus no one knows what causal links could underlie the statistical links. All this interest and effort betrays, I believe, the need to manage left-handers and the negative associations of left-handedness and the term "left".

What is, then, the scientific concept of left-handedness? First, there are three observations to be made. To begin with, there is more to handedness than simply "right or left"; for as long as there have been left-handers and right-handers, there have also been the ambidextrous. Still, three types of handedness, rather than two types, does not sufficiently reflect the nature of handedness. It was noted already in

[22] Göbel and Glöckler (2015, pp. 541–549) Translation from the German by the author.

a 1927 study that "dextrality and sinistrality are not opposed alternatives, but quantities capable of taking values of continuous intensity and passing one into the other."[23] In addition to that, some people prefer the right hand for some tasks and the left for other tasks in a consistent manner. Such a person has a strong hand preference, but might get the same "handedness quotient" on a test as someone who has a weak hand preference, while their hand preferences actually differ wildly. For example, one person might consistently write with the left-hand and throw a ball with the right hand, while another might use either hand for either task, and both might be considered ambidextrous or having "mixed handedness". One might therefore speak of two dimensions of hand preference, the right-left dimension and the strong-weak dimension. Strong hand preference can be anywhere on the right-left scale, but as the preference gets weaker, the possible right-left preference approaches the ambidextrous (or "mixed") middle to the point where having no hand preference allows only for being absolutely ambidextrous. Furthermore, a distinction is to be made between preference on the one hand and ability or proficiency on the other. Some people do not prefer the hand which has more strength or better motor skills. This applies in particular to people who have little or no hand preference despite one hand being stronger or more skilled than the other. However, the general correlation of hand preference and hand proficiency in non-right-handers is still a matter of debate.[24]

My second observation is that the hands are not the only lateralized part of the human body. Most people have a preference for foot, eye and ear, and this preference may or may not coincide with hand preference. About 10% of humans are left-handed, but approximately 20% are left-footed, 30% "left-eyed" and 40% "left-eared".[25] Some people have consistent side dominance, and this is more common among right-handers than left-handers, but many are cross-lateral.

Thirdly, as I noted above, measures of handedness result in a scale from left to right with no obvious divisions of people into different handedness groups. Usually the scale is either divided into three parts, left-handed, right-handed and mixed-handed/ambidextrous; or the left-handed and mixed-handed are grouped together as "non-right-handed". When that last label is used, one may be forgiven for the temptation to think of the non-right-handed group as "wrong-handed".

What these observations amount to is that any attempt to make the common sense (or pre-scientific) conception of left-handedness scientifically rigorous results in a diffuse picture of the laterality of the brain and the rest of the human body where left-handedness is merely a special case of the laterality of the body and the brain, and the classification of left-handedness is to a large extent arbitrary. Still, most researchers studying left-handedness choose to ignore these complexities and look for instruments to measure handedness that result in a neat division of right-handers, left-handers and the in-between group of the ambidextrous or mixed-handedness.

---

[23] Woo and Pearson (1927).

[24] See Bishop (1989); Corey et al. (2001) and Brown et al. (2004).

[25] McManus (2002, pp. 153–154). See also Porac and Coren (1981).

The lack of a clear criterion for left-handedness has led some to look for the "essence" of left-handedness: some necessary and sufficient conditions that determine who is left-handed once and for all. One place to look is the brain. We know that the right hemisphere of the brain controls the movement of the left side of the body and the left hemisphere the movement of the right side. Since Paul Broca's discovery in the 1860s, we also know that our ability to speak resides in some significant way in the left hemisphere.[26] In other words, most of us are left-hemisphere dominant for speech. This may seem to be the solution then: Right-handers have a dominant left-hemisphere and left-handers, conversely, have a dominant right-hemisphere. Unfortunately, this solution is too simple. If one hemisphere is "put to sleep" by an injection of sodium amytal (a barbiturate) into the carotid artery on one side, half of the body becomes paralyzed (the side opposite to the hemisphere) and many mental functions stop or are slowed down. This test, called the sodium amytal test, can be used to determine which hemisphere controls speech functioning. The test has showed that only about 20% of left-handers and 3–4% of right-handers are right-hemisphere dominant for speech.[27] This has been confirmed by more recent studies using transcranial Doppler ultrasonography.[28] The overall laterality of the brain does therefore not strictly determine handedness. Right-hemisphere dominance is clearly much more common in left-handers than right-handers, but, nonetheless, 80% of left-handers are left-hemisphere dominant for speech.

There have been many speculations that left-handedness is caused by some neurological failure or damage to the brain. As right-handedness is the norm, it is assumed that when everything goes "right" in the neurological development of a child, it turns out right-handed. If something goes wrong, it may turn out left-handed. Many scientists, among them psychologist Stanley Coren who I will discuss below, believe that some cases of left-handedness are pathological in this manner. Coren has claimed that left-handedness is a marker for a variety of diseases and abnormalities, arguing that if something goes wrong in the neurological development so that a person turns out left-handed, then it is very likely that something else is wrong too.[29] He even coined the word "alinormal" to describe those double deviants who are "left-handed but otherwise normal."

The exact causes of left-handedness (or human lateralization in general) remain poorly understood and the determination of left-handedness is still based on self-reported behaviour by way of questionnaires, which are sometimes supplemented with behavioural tests.

---

[26] McManus (2002, pp. 10–13).

[27] Coren (1992, pp. 100–102); Corballis (1983, pp. 67–69). For a fascinating discussion of brain lateralization, see McManus (2002, pp. 168–201).

[28] McManus (2002, pp. 200–201).

[29] See Chapter 9 "The Sign of the Left," in Coren (1992, pp. 153–164).

5.4 Do Left-Handers Die on Average Younger than Right-Handers?

From 1988 to 1994, at least three studies tried to find out whether left-handers die younger than right-handers. In this section I will look at these studies and discuss how the claim that left-handers die younger than right-handers was produced and circulate as scientific knowledge, and the power effects of this scientific work and the resistance that took place. First of all, the question seems to be quite straightforward. Statistics about life expectancy are familiar to most people, showing, for instance, that females can expect to live longer than males. One would think it should be quite simple to determine whether there is any difference in life expectancy between left-handers and right-handers. Still, it is not obvious why this question should be posed in the first place. One can see how biological as well as social and cultural differences can account for the different life expectancy of males and females, but such differences between left-handers and right-handers are much less obvious. As left-handedness is defined pre-scientifically, i.e. as a preference of using the left-hand for some salient tasks, there is nothing to indicate the possibility of a higher mortality risk for left-handers. Why would it occur to anyone that, say, if you prefer writing with your left hand you might expect to die younger than those who prefer using their right hand?

There are to two main reasons usually given for raising the question. First, some studies indicate that there are many more left-handers in younger age groups than the older ones. It has been reported that 13% of 20-year-olds are left handed, 5% of those in their 50s and virtually nobody of 80 or above.[30] The other reason is that left-handedness has been associated with all sorts of health risks such as alcoholism, smoking, illicit drug use, breast cancer and several neurological and immune disorders.[31] Left-handers may be more prone to accidents,[32] and left-handedness has been associated with other abnormalities, such as "dyslexia, stuttering, hyperactivity or attention-deficit disorder (ADD), autism, schizophrenia, tardive dyskinesia, mental retardation, epilepsy, early-onset Alzheimer's disease, neural tube defects, skeletal abnormalities, severe prematurity, cleft palate, and chromosomal abnormalities."[33]

And this list is still getting longer. It is striking how much of left-handedness research is concerned with various links between left-handedness and physical, mental, behavioral and social deviations. There has also been some, but much less, research conducted on the association of left-handedness with increased cognitive abilities and creativity.

---

[30] Halpern and Coren (1988).

[31] See in general London (1989); and in particular, association of left-handedness with alcoholism: London et al. (1985); with smoking: Harburg et al. (1978); with illicit drug use: Pretiab et al. (2012); with breast cancer: Titus-Ernstoff et al. (2000); with neurological disorders: Coren and Halpern (1991, p. 102); with immune disorders: Geschwind and Behan (1982).

[32] Coren (1989).

[33] London (1989, p. 1041).

Remember the pre-scientific definition of left-handedness: Left-handers prefer to use their left hand for writing and eating (or salient tasks of that sort). This sorts people quite neatly into left-handers and right-handers, clearly enough for everyday needs. But as I discussed in the previous section, when left-handers become the subject of study, it is not so clear anymore who is left-handed and who is not. In the studies which I shall look at, there is a strong tendency to define the subject so as to approach the pre-scientific human kind. This reveals yet again the concern with left-handers and left-handedness, as opposed to the more general concern with human laterality or cerebral asymmetry. In other words, scientists seem not so much concerned with finding the truth about various kinds and degrees of laterality, but with finding the truth about left-handers as a group or a kind of people.

Above I mentioned the claim that the proportion of left-handers is reduced in older age-groups. Such claims often refer to findings reported in a 1981 book by Clare Porac and Stanley Coren.[34] This observation can be explained in three ways:

(a) *The modification hypothesis*: There are fewer old left-handers because until recently many left-handers were forced to become right-handed. Younger left-handers are not confronted with the same pressure as the older ones were.
(b) *The elimination hypothesis*: There are fewer old left-handers because left-handers die younger than right-handers on average.
(c) *The bad sample hypothesis*: The numbers reported by Porac and Coren are wrong.

Stanley Coren and his co-authors are the main and possibly the sole source of evidence for the elimination hypothesis, that is, the scientific claim that left-handers die on average younger than right-handers. Of the three studies I will be concerned with here, two were conducted by Stanley Coren and Diane F. Halpern. Both are psychologists. The third study was conducted by Marcel E. Salive and Jack M. Guralnik, who were with the Epidemiology, Demography and Biometry Program at the National Institute on Aging, National Institutes of Health in Maryland, and Robert J. Glynn, who is at the Harvard Medical School.

The first study by Coren and Halpern used a rather unorthodox source of data, namely The *Baseball Encyclopedia*.[35] As Coren put it in an interview: "What we needed was a set of individuals, all of known handedness and all already dead. And that's not easy to find." Coren and Halpern used records for 2271 major-league baseball players, excluding those baseball players whose handedness was ambiguous. The definition of left-handedness used here is that a player is classified as left-handed if he uses left hand for both throwing and batting throughout his career. The result of the study was that the mean age at death turned out to be approximately 8 months lower for left-handers than right-handers.

In 1988, Stanley Coren and Diane F. Halpern announced the results of their study in a letter to *Nature*.[36] Although the apparent difference in mean age at death is not

[34] Porac and Coren (1981).

[35] Halpern and Coren (1988).

[36] Ibid.

great, the results made the headlines of newspapers; radio and TV networks picked up the story and, of course, science magazines reported and discussed the findings. Articles on Coren's study appeared in as diverse publications as *Weight Watchers Magazine*, *Parents* (entitled "Attention, Parents of Left-Handers") and *Chemical and Engineering News*.[37] Although the difference in life expectancy is not great, this new bit of scientific knowledge was quickly circulated in various media.

In the scientific magazines, letters and articles poured in criticizing Coren and Halpern's study. Many letters pointed out that the data are not that representative of the normal population. One critic noted that "left-handed women cricketers need not worry unduly."[38] Other critics appealed to the modification hypothesis, that there are fewer left-handers to be found in older age groups simply because so many of them had been forced to become right-handed. Most of the criticism was concerned directly with Coren and Halpern's classification of handedness and their statistical reasoning. Max G. Anderson, described as "a left-handed number cruncher with the Canadian Statistical Analysis Service in Vancouver" applied different statistical methods and a slightly different way of classifying left-handed players, thereby including some ambidextrous players in the left-handed group. He used a more recent edition of a sports encyclopedia, and came to the opposite conclusion: left-handers live longer than right-handers.[39] Further criticism came from a study conducted by Elizabeth Wood at the California Institute of Technology. Like Anderson, she failed to replicate the results using a more recent edition of the Baseball Encyclopedia. She also evaluated Halpern and Coren's statistics and found that

> even if there were no difference in the mortality distributions of left-handers and right-handers, a difference in the cumulative survival fraction larger than that presented by Halpern and Coren would arise by chance in about 98 out of 100 samples of the size presented here.[40]

The conclusion drawn from her study was that in that limited sample "there is no statistically significant difference between mortalities of left-handed and right-handed baseball players."[41]

In his book *The Left-Hander Syndrome*, Stanley Coren makes some interesting remarks on this criticism. Throughout his book, whose target group appears to be non-scientist left-handers, he gives a rather idealistic picture of scientific practice: he is the rigorous scientist working in his laboratory and discovering things about his subject in accordance with strict scientific methods. Coren frequently talks about his "laboratory" and even "the experimental method," when his research is almost exclusively based on statistical reasoning and statistical data. It is perhaps not very surprising that he gets somewhat annoyed when that self-image is broken and he

[37] See Coren (1992, p. 213); Duarte, (1990, p. 16); and Reese (1988, p. 56).

[38] W. W. Nichols of John Radcliffe Hospital in Oxford, quoted in Reese (1988, p. 56).

[39] Weiss (1989, p. 180); see also Anderson (1989, p. 112).

[40] Wood (1988).

[41] Ibid.

gets criticism not from fellow scientists but from his subjects, the left-handers. Coren does not respond to their criticism, but questions their motives:

> Perhaps it was predictable that the two major comments on our findings were either authored or directed by left-handers and that none of the attacks was initiated by psychologists or neuropsychologists who knew the data that had led us to conduct the study in the first place. The initial comment came from E. K. Wood, who is described in one article as being "neither left-handed nor a scientist" but whose analyses were directed by E. Sterl Phinney, a left-handed professor of theoretical astrophysics from the California Institute of Technology. Another broadside came from Max Anderson, a left-handed free-lance statistician, working out of his home in Vancouver. Both comments pointed out some of the problems associated with working with archival data.[42]

Coren then goes on to explain what archival data is and the problems such data poses for this type of research. He notes that "when using archival data, the strength and reliability of results that the researcher gets depend [...] upon the definition of handedness and the specific statistical tests used to determine the significance of the findings."[43] Coren had already made a comment to this effect in a lengthy review published in the *Psychological Bulletin* in 1991, only a few months before his book was published: "It appears that the statistical significance of these results varies as function of the specific statistical analyses used and as a function of the definition of handedness employed (e.g., Anderson 1989; Wood 1988)."[44] What he does not say is that all attempts to replicate his results from the same or similar sources (in particular sports encyclopedias) had failed. The statistical methods he chose and the definition he employed happened to be the only ones to show that there is any considerable difference in the longevity of lefties and righties at all.

As Coren grudgingly pointed out, much of the criticism he faced came directly or indirectly from left-handers who were not psychologists, but well-versed in statistical methods. His scientific authority was challenged by his research subjects, leading to a struggle that took place outside his scientific discipline but still on scientific grounds. Coren's reaction to the criticism, as could be seen in the quote from his book above, was to reassert his scientific authority and dismiss his critics with ad hominem attacks: they are outsiders from his discipline; they are left-handers and hence biased; one critic is not a scientist and another critic is working out of his home.

Stanley Coren and Diane Halpern decided to conduct another study, based on a more reliable set of data. They obtained death certificates from two counties in southern California.[45] Two thousand questionnaires concerning the handedness of the deceased family member were sent to the listed next of kin, which resulted in 987 usable cases. They asked which hand the deceased used for writing, drawing and throwing a ball. If the deceased used the right hand for all the tasks, he or she was classified as right-handed, otherwise as left-handed. Coren notes that this clas-

[42] Coren (1992, p. 214).

[43] Ibid.

[44] Coren and Halpern (1991, p. 214). The references are to Anderson (1989) and Wood (1988).

[45] Coren and Halpern (1991).

sification includes ambidextrous in the left-handed group. He labels the group left-handed, although "non-right-handed" would be a more appropriate label. Coren surprisingly remarks that he and Halpern "were insisting that the person be strongly right-handed to be counted as a right-hander, much as we did with the baseball player data."[46] On the contrary, the baseball player study classified left-handers as strongly left-handed, not as left-handed and ambidextrous.

In 1991, 3 years after the baseball study, Coren and Halpern published the results of their Californian study in *The New England Journal of Medicine*.[47] Their result was that left-handers die on average 9 years earlier than right-handers.

This second study of Coren and Halpern did not get any warmer reception than the first one and was widely criticized. Again the statistical methods were found faulty. Tricia Hartge, an epidemiologist at the National Institutes of Health, remarked that the results "mean absolutely nothing" because Coren and Halpern never found out the number of the left- and right-handed in those two counties, nor their age distribution. Actuaries ridiculed the study. The American Academy of Actuaries issued a statement calling the statistical methods faulty and Richard Labombarde, an actuary with the private firm of Milliman and Robertson in Washington, is quoted in the New Scientist saying that the research reminded him of methods used by actuaries 200 or 300 years ago.[48]

The strongest criticism comes from the epidemiologist who conducted the third study I will discuss here, Marcel E. Salive. He noted that if the age distribution was anywhere close to what is reported in Porac and Coren's 1981 book, that is, if left-handers are on average so much younger than right-handers, and if the death rate is the same for the two groups, the difference in mean age at death should be even greater than what Coren and Halpern found in their California study. In other words, the age difference may be the cause and not the result of difference in mean age at death. Therefore, the difference in mean age at death does not by itself indicate what the difference in life expectancy may be, or whether there is a difference in life expectancy.[49] Chris McManus directs quite the same criticism at the study in his *Right Hand, Left Hand*, discussing the study in a chapter called "Vulgar Errors."[50] Noting that left-handers are on average younger than right-handers (since there are more of them now than a few decades ago), he points out:

> People who read Harry Potter books are younger than those who do not. Ask the relatives of a group of recently deceased people whether their loved one had read Harry Potter and inevitably one will find a younger age at death in the Harry Potter enthusiasts, but that is only because Harry Potter readers are younger overall. There is no need for a government health warning on the cover of Harry Potter books.[51]

---

[46] Coren (1992, p. 218).

[47] Coren and Halpern (1991).

[48] Charles (1991).

[49] Salive et al. (1993).

[50] Chris McManus (2002, pp. 292–293).

[51] Ibid., p. 293.

Despite all this criticism, the findings were no less widely publicized than the earlier findings. Again the scientific knowledge had been produced and put into circulation, whatever one may think about its reliability or of the way it was produced.

I do not know to what extent this criticism came from left-handers. Resistance from left-handers did, however, take a non-scientific form when some left-handers started phoning in death threats.[52] This reaction, this apparent anger of left-handers, may not be a very strong move strategically, but it does show that this production of scientific knowledge was having a considerable effect on the subjects of the study. We can call these effects power effects, for they are obviously affecting the behavior and self-image of left-handers. The philosophical point to note here is that the conflict is not about the application of science, but about certain scientific knowledge and its immediate effects. The resistance to this knowledge production is not concerned with a risky application or misuse of scientific results. It is concerned with the knowledge production itself and its role in biopolitics in Foucault's sense.

After this second study and the criticism it got, science journals such as *Nature* and *New Scientist* observed that in order to produce the knowledge in question, other methods are needed.[53] The best way to find out whether left-handers die on average younger than right-handers is to conduct a cohort study, that is, to take a group of right-handers and a group of left-handers and keep track of how many die from each group over a considerable length of time.

The third study I will look at was conducted by epidemiologists Marcel E. Salive, Jack M. Guralnik and Robert J. Glynn. They analyzed data from 3809 people aged 65 and older and recorded the mortality rate over 6 years. Subjects were classified "as right-handed if they reported using their right hand for both writing and cutting with scissors and as left-handed if they used their left hand or either hand for either task."[54] Salive classifies the subjects as left-handed if they are not strongly right-handed. As he remarks himself, this classification principle is comparable to that used in Coren and Halpern's California study and in Porac and Coren's 1981 book. It is, however, different from the classification used in Halpern and Coren's Baseball Encyclopedia study, where a player is classified as left-handed if he uses left hand for both throwing and batting throughout his career. Salive's study found left-handedness not related to age, contrary to Porac and Coren's 1981 book—the study which initiated this line of research. The 6 year mortality rate turned out to be almost the same for the right-handed and the non-right-handed: Over the 6 years (1982–1988) 33.8% of the left-handers died and 32.2% of the right-handers. The difference is not statistically significant. These results, therefore, support the null hypothesis that there is no difference in the mortality rate of right-handers and left-handers.

[52] Holden (1991).

[53] Pool (1991); and Charles (1991).

[54] Salive et al. (1993, p. 265).

Coren and Halpern's reaction to this study was to claim:

> It is a basic principle of the experimental method that failures to reject the null hypothesis cannot be used to cancel statistically significant results. There are many ways to obtain results that are not statistically significant, including sloppy research and low statistical power.[55]

Salive's study, they say, falls under the latter, namely low statistical power. It is worth noting, however, that Salive's sample was almost four times as big as Coren and Halpern's sample in their California study, and nearly 70% larger than the sample in their baseball study. Apart from Coren and Halpern's criticism, the reactions to Salive's study were minimal and it received no publicity in the popular media.

5.5 Conclusion

Despite the fact that the only evidence for the claim that left-handers die younger than right-handers are the two studies by Coren and Halpern, and despite the criticism, the presence of studies failing to replicate the results and the absence of studies replicating the results; it is still widely believed that left-handers do on average die younger than right-handers.[56] This has to do with how scientific knowledge is produced, circulated, diffused and consumed. The results of the two positive studies were presented not only in scientific journals, but also in all sorts of popular and semi-popular magazines (like *Weight Watchers Magazine*, *Parents* and *Chemical and Engineering News*). The results were also discussed in the mass media, culminating perhaps in Stanley Coren's appearance on an American television talk show (including a confrontation with a proud but upset left-hander and a cheering/booing studio audience). In contrast, the cohort study which found no difference in life expectancy between right-handers and left-handers got hardly any attention in the media at all.

This case of biopolitics shows how knowledge about the longevity of left-handers was produced and circulated, and what power-effects it has had, as well as the resistance of left-handers, namely criticism at the scientific level and down to death threats. The knowledge in question is clearly within the realm of biopolitics, as it is concerned with the statistical study of life expectancy of left-handers. The motivation for this knowledge production is not directly to help manage or govern the population of left-handers, but it is squarely within a research context which has its roots in the management of left-handers, in particular left-handed pupils, and in the need to know the nature of left-handers as a certain sort of abnormal people.

---

[55] Coren and Halpern (1993, p. 238).

[56] See for example Hannah Barnes (2013).

References

Anderson, Max G. 1989. Lateral preference and longevity. *Nature* 341 (6238): 112.
Barnes, Hannah. 2013. Do left-handed people really die young? *BBC News, Magazine* (7 September 2013), http://www.bbc.com/news/magazine-23988352. Accessed 18 Dec 2015.
Binnie Dawson, John L.M. 1977. An anthropological perspective on the evolution and lateralization of the brain. *Annals of the New York Academy of Sciences* 299.
Bishop, D.V.M. 1989. Does hand proficiency determine hand preference? *British Journal of Psychology* 80: 191–199.
Blau, Abram. 1946. *The master hand: A study of the origin and meaning of right and left sidedness and its relation to personality and language*. New York: The American Orthopsychiatric Association Inc.
Brown, Susan G., Eric A. Roy, Linda E. Rohr, Benjamin R. Snider, and Pamela J. Bryden. 2004. Preference and performance measures of handedness. *Brain and Cognition* 55: 283–285.
Burt, Cyril. 1937. *The backward child*. New York: Appleton.
Charles, Dan. 1991, April 27. Left-handers don't die young after all. *New Scientist* 1766: 21.
Cholmeley, Roger James, ed. 1919. *The Idylls of Theocritus*. London: G. Bell & Sons.
Corballis, Michael. 1983. *Human laterality*. New York: Academic Press.
Corballis, Michael C. 1991. *The Lopsided Ape*. Oxford/New York: Oxford University Press.
Coren, Stanley. 1989, August. Left-handedness and accident-related injury risk. *American Journal of Public Health* 79 (8): 1040–1041.
———. 1992. *The left-hander syndrome: The causes and consequences of left-handedness*. New York: The Free Press.
Coren, Stanley, and Diane Halpern. 1991. Left-handedness: A marker for decreased survival fitness. *Psychological Bulletin* 109: 90–106.
———. 1993. Left-handedness and life span: A reply to Harris. *Psychological Bulletin* 114: 235–241.
Corey, David M., Megan M. Hurley, and Anne L. Foundas. 2001. Right and left handedness defined: A multivariate approach using hand preference and hand performance measures. *Neuropsychiatry, Neuropsychology, and Behavioral Neurology* 14 (3): 144–152.
Duarte, M.Y. 1990, May. Attention, parents of left-handers. *Parents*: 16.
Foucault, Michel. 1980. *Power/knowledge: Selected interviews and other writings 1972–1977*. New York: Pantheon Books.
———. 1990. *The history of sexuality*. New York: Vintage Books.
Gay, John. 2005. *Fables: With a Memoir by Austin Dobson*. Whitefish: Kessinger Publishing.
Geschwind, Norman, and Peter Behan. 1982. Left-handedness: Association with immune disease, migraine and developmental learning disorder. *Proceedings of the National Academy of Sciences* 79: 5097–5100.
Göbel, Wolfgang, and Michaela Glöckler. 2015. *Kindersprechstunde: ein medizinisch-pädagogischer Ratgeber*. 19th ed. Stuttgart: Urachhaus.
Hacking, Ian. 1999. *The social construction of what?* Cambridge, MA: Harvard University Press.
Hacking, Ian. 2002. *Historical Ontology*, 99–114. Cambridge, MA: Harvard University Press.
Hacking, Ian. 2007. Kinds of people: Moving targets. *Proceedings of the British Academy* 151: 285–318.
Halpern, Diane F., and Stanley Coren. 1988, May 19. Do right-handers live longer? *Nature* 333 (6170): 213.
Harburg, Ernest, Anna Feldstein, and James Papsdorf. 1978. Handedness and smoking. *Perceptual and Motor Skills* 47: 1171–1174.
Harris, Lauren Julius. 1980. Left-handedness: Early theories, facts and fancies. In *Neuropsychology of left-handedness*, ed. Jeannine Herron, 3–78. New York: Academic Press.

Hertz, Robert. 2013. The pre-eminence of the right hand. A study in religious polarity. *HAU: Journal of Ethnographic Theory* 3 (2): 335–357.

Holden, Constance. 1991, May 17. Slugging it out over left-handed mortality. *Science* 252 (5008): 916.

Liddell, Henry George, and Robert Scott. 1992. *Intermediate Greek-English Lexicon*. 7th ed. Oxford: Oxford University Press.

London, Wayne P. 1989. Left-handedness and life expectancy. *Perceptual and Motor Skills* 68: 1040–1042.

London, Wayne P., Priscilla Kibbee, and Laurent Holt. 1985, September. Handedness and alcoholism. *The Journal of Nervous and Mental Disease* 173 (9): 570–572.

McManus, Chris. 2002. *Right hand, left hand: The origins of asymmetry in brains, bodies, atoms and cultures*. Cambridge, MA: Harvard University Press.

McMullin, Walter G. 1914, December. What shall we do with Left-handed pupils? A Symposium conducted by Walter G. McMullin. *Teacher* 18: 331–338.

Pool, Robert. 1991, April 18. Can lefties study be right? *Nature* 350 (6319): 545.

Porac, Clare, and Stanley Coren. 1981. *Lateral preferences and human behavior*. New York: Springer Verlag.

Pretiab, Antonio, Ileana Usaia, Elisa Pintusa, Cinzia Sardua, Donatella Rita Petrettoa, and Carmelo Masala. 2012. Left-handedness is statistically linked to lifetime experimentation with illicit drugs. *Laterality: Asymmetries of Body, Brain and Cognition* 17 (3): 318–339.

Reese, Kenneth M. 1988, September 5. Do right-handers live longer than left-handers? *Chemical and Engineering News* 66 (36): 56.

Salive, Marcel E., Jack M. Guralnik, and Robert J. Glynn. 1993, February. Left-handedness and mortality. *American Journal of Public Health* 83 (2): 265–267.

Titus-Ernstoff, Linda, Polly A. Newcomb, Kathleen M. Egan, et al. 2000. Left-handedness in relation to breast cancer risk in postmenopausal women. *Epidemiology* 11: 181–184.

Uomini, Natalie T. 2009. The prehistory of handedness: Archaeological data and comparative ethology. *Journal of Human Evolution* 57 (4): 411–419.

Weiss, Rick. 1989, September 16. Lefties and longevity: Look again. *Science News* 136 (12): 180.

Woo, T.L. (Ding-Liang), and Karl Pearson. 1927. Dextrality and sinistrality of hand and eye. *Biometrika* 19: 199.

Wood, Elizabeth K. 1988, September 15. Less sinister statistics from baseball records. *Nature* 335 (6187): 212.

Chapter 6
The "Me Molecule"

Judit Sándor

Abstract The cells, cell lines, embryos, and gametes that are harvested, isolated, purified, fertilized, frozen, stored, and defrosted, tested, diagnosed, examined and even "edited" in laboratories. As a consequence of this multiple form of intervention human cells and molecules have recently become the subject of vibrant political and ethical debates and targets of legal policymaking. In this scientific and political discourse a new phenomenon has emerged, which was coined as 'molecularization' by Nikolas Rose in his book, *The Politics of Life Itself.* For Rose, molecularization is one of the most important characteristics of contemporary biopolitics. When Foucault elaborated his ideas about biopolitics and subjectivity, he still focused on the body as a whole, and the biopolitical control over it, and not on the parts or fragments of the body (Lemke T. Biopolitics: an advanced introduction, New York University Press, New York, 2011). Today, however, not only scientists, but also regulators view humans more and more on the molecular level by developing special scientific terminology and legal norms in the fields of genetic research and testing, stem cell research, and even assisted reproduction. This phenomenon has appeared in several domains of biotechnology, which I will discuss by highlighting the consequences of this new scientific gaze. In this chapter, therefore, I would like to discuss the practical implications of molecularization, and explore how the new biopolitical thinking determines legal structures and public discussion in the light of the current advances in biotechnology. My goal is to incite a legal dogmatic dispute and to encourage a brainstorming that incorporates an element of criticism in the process of placing biological concepts in a new context. I will examine the relevance of molecularization in selected fields, such as genetic research, biotechnological inventions, sperm donation, mitochondrial donation and surrogacy.

Keywords Biolaw • Biotechnology • Rose • Human rights • HGP • Patents • Surrogacy

---

J. Sándor (✉)
Central European University, Budapest, Hungary
e-mail: sandorj@ceu.edu

P. Kakuk (ed.), *Bioethics and Biopolitics*, Advancing Global Bioethics 8,
DOI 10.1007/978-3-319-66249-7_6

6.1 Introduction

One of the basic questions raised by biotechnology law is what norms can be applied to the human body, its parts, cells, molecules, and specifically to human DNA. The field of genetics examines hereditary properties and factors that may be influenced to a limited extent, and therefore, any misuse of our genetic data may have serious adverse consequences. Human DNA is not only an important tool to study hereditary conditions but to the public it has also manifold symbolic meanings. It represents the link to ancestors, it refers to paternity, in some cases it reveals the identity (missing persons, stolen babies), and it serves as important evidence in judicial cases (criminal cases, paternity cases etc.). As a consequence, genetic data has become subject to higher scrutiny than ordinary health care data (Jasanoff 2011). At the beginning even some exaggerated claims were raised to consider genetic data as an entirely new phenomenon but by now the public and the legislature applies more nuanced approaches when genetic data and tests are concerned. For instance when genetic data is being used for the purposes of diagnosis or therapy, society shows a more liberal approach while in case of commercial uses of human stem cells and DNA more prohibitory stance can be observed.

In my view this suggests that we cannot deal with regulatory issues of human cells and DNA as if they represent a monolithic ethical-legal issue and which simply extends the legal norms from the human body to human cells. There are obviously lots of similarities between the protection of the human body and the protection of the human DNA. For example, both the body and the DNA may be carriers of important personal information; therefore, in the field of data protection norms, it is inevitable that there is an overlap. However, once we consider human rights norms that aim to protect individuals against suffering and abuse, the differences between the two fields are striking.

In order to discuss this highly complex and abstract legal policy field, several sub questions have to be addressed. The first thing which has to be asked is what is the legal status of the human cells, gametes, DNA? Should we use a case-by-case, field-by-field, use-by-use approach? Then the further question is to what extent non-commodification shall apply in the field of molecular "products", donation of cells and tissues. Then I will examine the evolution of a personalized cell and legal advantages and disadvantages of this phenomenon followed by the conclusion on a possible legal policy that better captures molecularization. My objective is to investigate an issue that the majority of decisions-makers and legislators have so far basically neglected: do human rights, fundamental rights applicable to persons may automatically be applied to the field of human specimens, human cells, and molecules?

I would like to emphasize that objectification and commodification of persons have been a long recognized problem, much earlier than the emerging possibility of studying humans on the molecular level. Martha Nussbaum (Nussbaum 1995) developed seven different concepts involved in objectification of a person: instrumentality, denial of autonomy, identification with inertness, fungibility, and violability, ownership of a person like an object, and denial of a person's subjectivity. The identification of the body as property directs attention, of all things, to inequality, rather than to the joint experience that makes all humans vulnerable through birth, illness, aging and mortality, even though our shared bodily fate and transience creates a sense of mutual interdependence in us.

What is fascinating, and also quite confusing, is that law now interprets the human body and its parts differently in various biotechnological contexts. Human body parts, such as organs or blood, are usually covered by the laws on transplantation and transfusion. Other, smaller body components, such as cells, genes, or proteins, have become relevant as objects of legal regulation especially after the Human Genome Project. That is when molecularization has started to influence legal policies (Laurie 2002).

As a result of the debates unfolding in the fields of ethics and law after the Human Genome Program, genetics served as a catalyst to biomedical law, as it called into question several, previously taken for granted norms and principles, or at least forced legislators to conceive new interpretations. This gives rise to several further questions: can the individual's self-determination and control over one's own genetic samples even be ensured? And the other main question also emerging: in what ways can the information derived in such a manner be used in various research projects? It appears that human tissues removed during surgeries, which was only considered hazardous waste previously, is increasingly seen as having research and even commercial significance. After all, even from a vial of blood it is possible to determine a large amount of medical and other, personal information, not to mention DNA, which is particularly valuable from a research point of view, as well as human organs and tissues.

These dimensions of the vulnerability of privacy rights (Laurie 2002) have already appeared in several legal cases as well. Instead of an extended historical overview, I will only mention a few, markedly different legal perceptions here, by way of examples.

6.2 Things, Persons or Particular Subjects to Biotechnology Law?

According to Roman law, there can be no ownership of body parts of free human beings (Under the Roman law a living human body belongs to the category of 'persons' and not to 'things' (*res*) and even the dead body is considered as *res extra commercium*) (Agamben 1998). This statement can be achieved by applying legal dogmatic approach or a simple logic, by saying that in the case of the human body, the owner and the owned are identical, and therefore, the concept of ownership cannot be interpreted in such a situation. In the opinion of Locke, however we can have ownership over our own bodies. "Every man has a property in his own person" (Locke 1689). But this is not meant property in the true material sense; in other words, it is not the physical body to which ownership applies. In a similar vein, Jeremy Waldron states that "humans, then do not have creators' rights over their bodies. But they can be regarded in the strong sense as the creators of their own action." Although creator's rights is a kind of vague term still it captures the difference (Waldron 2006) between the right to dispose of and right to decide on the human body. In the current biotech law others (for example, Bovenberg) suggested that the freedom to research the human genome was akin to the freedom of navigation, and quoted Grotius, the famous scholar of natural law, in drawing up an analogy between the status of the high seas and the legal status of the human genome. Of course, these various, partly philosophical and partly legal concepts have numerous practical consequences, from the limits to the freedom of research to the patentability of the human body. In addition, there also exist mixed legal solutions side by side where the various areas of law offer different evaluations of the legal status of human tissue. For example, from the point of view of criminal law, stealing a biological specimen constitutes theft, even though from a civil law point of view the same specimen could not be the owned as property. At the same time, in the case of organs, tissue and blood samples, it is, indeed, a very important consideration that they are of human origin, meaning that at one point there was or even at the present there is a human being behind them. It was a question raised in numerous criminal cases whether a person has property in his or her waste products, human excrement and urine. This is of particular significance when the law enforcement authorities are able to obtain data that can serve as evidence in a criminal case.

6.3 Can Researchers and Biotechnological Companies Gain Profit on Human Cells, Tissues, DNA?

In order to come to the issue of patents on human DNA, it should also be clarified first to what extent the personality can be seen as inherent with a biological specimen, tissue or genetic sample of human origin, and what significance can be attributed to who obtained these and in what way. Further, in each case the additional question must also be examined whether it is human dignity or the right of ownership that should be protected in the given situation. In other words, should we develop our arguments on personality rights or on property rights? (Ackerly 2008).

We know that after a certain stage of the biotechnological procedure, human tissues and DNA also have commercial value, and therefore, they may be the subject of property rights. According to Beyleveld and Brownsword, the current ways in which tissues are used create a marked pressure of commodification (Beyleveld and Brownsword 2000). Although their work originally covered the field of personality and privacy rights, their research shifted to discuss also the problems of commodification. We cannot ignore the fact either that at the end of the biotechnological procedure there may also be a pharmaceutical product, a hormone preparation, or an invention the origin of which was an organ or tissue sample, or a cell or stem cell line, cell sample of human origin, this may easily lead to bad legal conclusions.

In 1995, still before the major biobank projects were launched, hospitalized patients in the UK were asked who they thought the removed organ or tissue should belong to (Weir and Olick 2004). Twenty-seven percent of the respondents were of the opinion that the hospital can retain the organ or tissue removed in the course of the surgery. Pathological laboratories were also identified by a high number of respondents (20%), and further, many believed that there can be no property rights over organs and tissues thus removed—that is, these are to be considered as *res nullius*.

Personalization of human biological samples can be seen as byproduct of biotechnology itself (Sándor 2012). Scientists named and marked new cell lines and discoveries that often referred to the name of the actual person from whom the sample has been originated. The most well-known example is the HeLa cell line. The generation of cell lines is of major purport not only in genetics and molecular biology, but also in immunology. These play a role in the treatment of cancer and other serious illnesses, as well as the identification of the causal mechanisms of these illnesses. In 1981, Hideaki Hagiwara, a young researcher in California proposed that cancer cells from his mother's cervix be used in an important research project. They subsequently succeeded in producing a cell line from the tissue sample, which held out high hopes of not only generating valuable intellectual property, but also of significant actual benefit in therapy. A dispute arose, however, between the research team and the researchers supplying the tissue sample concerning the exploitation of the invention. Hagiwara wanted his mother's name to survive on the tissue sample (but at least the Hagiwara name to be continued as part of the name used in Asia). Eventually, this was the result of the out-of-court settlement, in contrast with the most famous Moore case.

The decision in the *Moore v. The Regents of the University of California* case had an enormous influence worldwide. John Moore was diagnosed with hairy cell leukemia, a rare and very serious disease. Because the patient's father was a physician himself, he suggested to his son to seek treatment at the UCLA Medical Center in Los Angeles, which had a good reputation in treating this illness. Indeed, the treatment proved to be successful. Moore regularly had to travel from Seattle to Los Angeles to various treatment sessions, in the course of which different tissue samples were taken from his body on several occasions, including bone marrow, blood, sperm and skin tissues. In each case, Moore was told that these examinations were necessary for his post-operative care to determine whether the therapy was effective. He did not know that the research team was motivated by interests other than his treatment. In connection with the case, the ethical question emerged to what extent and depth must a patient be informed that physicians are motivated not only by the interests of therapy, but also research, and in the given case also by commercial interests. If the patient had received such information, he could have obtained a second medical opinion and clarified which elements of the treatment actually serve his therapy-related objectives. Or, it may have transpired that it was no longer necessary for him to travel such large distances for the purposes of post-treatment. The work of the research team in Los Angeles finally led to a patent, which otherwise generated a large amount of income for the university, and an important cell line, the "Mo cell line" was also produced from this work in 1984.

The surprising twist when he was asked to sign a consent form of retroactive effect under which he would have granted to the university exclusive rights in the cell lines and biological samples removed from him in the course of the research and his treatment, as well as all rights that could be derived from it. Researching the legal literature and the patents, Moore's lawyer realized that the Mo-cell line was created during this time, and it was probably for the already existing patent that Moore's retroactive consent would be needed. In his statement of claim, Moore wanted to participate in the income; that is, he demanded a share from the profit generated by way of the patent. The court rejected his claim: the legal reasoning used relied on the analogy of transformation, according to which, similarly to works of art, the significance of the raw material is negligible compared with the amount of work invested, and therefore, Moore's claim for ownership was deemed unfounded. At the same time, the court also found a breach of the fiduciary duties. According to this, it cannot be assumed that the contractual relationship between a physician and a patient is entirely symmetric, since the patient is in a vulnerable position: and this indicates more responsibility on the side of the physician to disclose biases and the conflict of interests. Based on this case, informed consent now needs to be specified in a way to communicate also the purpose of the tissue removal. The Moore case has since been cited by many, whenever issues related to the use and ownership of the human body and tissues arise. The case showed that the relevance of human tissues, cell lines require innovative legal policy as informed consent is no longer restricted to the medical interventions on the body. Cells, molecules, and tissue also matter.

6.4 Molecularization and Patentability

Another significant step in the development molecularization was related to the increasing number of biotechnology patents. Patents in genes have generated fierce debates ever since the famous *Diamond v. Chakrabarty* case. In this case, the U.S. Supreme Court ruled in 1980 that genetically modified bacteria were patentable, in contrast to the opinion of the patent officers earlier acting in the case who had rejected the claim. By the beginning of this millennium, as many as 6000 patents were registered for genes identified in living organisms. Patents in genes have since received much criticism based on legal and ethical considerations. Some of the counterarguments can only be applied to human biotechnology, while others are also valid for genetically modified organisms used, for example, in the agricultural sector.

The political relevance of the DNA was manifested in 1997, in the Universal Declaration on the Human Genome and Human Rights adopted by UNESCO. This declaration summarized the generally accepted principles of bioethics, such as respect for human dignity, the ban of financial gain from the use of tissues of human origin as well as the sharing of benefits and the advancement of scientific progress. As regards the ownership and patentability of human genes, the UNESCO has dealt with this issue already in two declarations.

The World Medical Association's Declaration of Helsinki considered the risk-benefit analysis of research in medical biology, the examination of the necessity of the research project, the protection of confidentiality and trust, as well as the observance of the requirement of informed consent as important points.

The Oviedo Convention of the Council of Europe on human rights and biomedicine, imposing restrictions on biomedical research and therapy for human rights purposes, builds on the protection of human dignity and integrity, and has reinforced the *primacy of the human being* over the interests of science and society (Articles 1 and 2). Despite the fact that, on the level of general principles, signatories were striving for uniformity, the Oviedo Convention reflects the diversity of European positions related to biomedical research. Respect for local interpretations can be found in the Explanatory Report (Section 18), providing that the meaning of "everyone", which is the subject of the respect for human dignity alongside with the phrase "human beings," must be defined in the national legislation. The convention partly leaves open the question concerning research on human embryonic stem cells in Article 18 (1): "Where the law allows research on embryos in vitro, it shall ensure adequate protection of the embryo."

Apart from the growing number of international, regional and national documents on the human genome, court cases also encompassed several new aspects of this field of research. It is enough to take a brief look at the Myriad case to see that judges took significant effort to interpret the meaning of the human DNA, as well as difference in the natural and artificial elements in genetic research.

One can find numerous examples of incorporating tissues and molecules in the contemporary biotechnology law. For instance, the *Regulation (EC) No 1394/2007 of the European Parliament and of the Council of 13 November 2007 on advanced therapy medicinal products and amending Directive 2001/83/EC and regulation (EC) No 726/2004* is dealing with a new domain of biotechnology. It is *lex specialis* which means that it provides a specific and additional legal framework to the already established norms of clinical trials. The regulation recognizes the new scientific progress which involves for instance in gene therapy, somatic cell therapy, tissue engineering (Zimmer et al. 2015).

Recital 13 of the Regulation claims that advanced therapy medicinal products should be subject to the same regulatory principles are other types of biotechnology medicinal products. Furthermore, the Regulations (in Recital 14) states that the regulation should also not derogate from the basic principles laid down in Directive 2004/23/EC. As it follows donation of human cells or tissues, principles such as anonymity of both donor and recipient, altruism of the donor and solidarity between donor and recipient should be respected.

6.5 What Is the Role of Human Rights in Molecularization?

As Hans Joas states, "The triumphal march of human rights is undoubtedly one of the great success stories in the realm of values and norms" (Joas 2013). After the second world war human rights norms developed to protect "equal and inalienable rights of all members of the human family". It is clear from the early human rights texts that these declarations, conventions and covenants aimed to cover "everyone". The inclusion of "everyone" referred to the imperative to provide protection to people "without distinction of any kind, such as race, colour, sex, language, religion, political or other opinion, national or social origin, property, birth or other status." Furthermore, discrimination (Duster 2003) was also not allowed on the basis of "the political, jurisdictional or international status of the country or territory to which a person belongs, whether it be independent, trust, non-self-governing or under any other limitation of sovereignty." After the Second World War when these norms were developed the main social evil in the field of biomedicine was the human experimentation which occurred before and during the Second World War and was subject also of the Nuremberg trial. But one cannot find any reference or even a vague hint to the intention that these early conventions were applicable to cover cells, tissues of human origins. Removal of a tissue from a human body during the course of an operation and then to use it covertly for research purposes was not regarded as a human rights issue, not even as a legal issue. When HeLA cell lines were created from the tissue removed from Henrietta Lack's body John Hopkins Hospital did not violate a law or an established ethical norm. The new phenomenon was the molecularization as described by Nikolas Rose which urged the need to reflect on the increasing number of research, interventions which were conducted not on the human person but only a blood sample, or DNA, tissue, or gametes. The "emancipation" of these bodily materials was stimulated by international norms, such as the Oviedo Convention and the first two relevant UNESCO Declarations adopted in 1997 and 2003.

6.6 The Role of the Human Genome Project in the Process of Molecularization

All these developments should be seen as a legal progress in the sense that human persons now should be protected also when research is being conducted on their DNA and gametes, and cells. I believe the confusion came with the fact that international community, patients, doctors, DNA donors may disagree on the question to what extent should we use these samples as parts of the individual, when can we see these sources as common research tools who should financially benefit from the fruits of the research. To what extent individual can use samples, cells, gametes to meet his/her very personal needs even when this need is not recognize by the community, furthermore especially in the field of assisted reproduction, gamete donation and embryo selection how can double emancipation problems be settled? I refer to the problem of double emancipation when the legislator protects embryos, sometimes even fertilized ovum because of their potentiality to develop a human fetus and then a person but in the same time gamete owner, people who contributed to the creation of the human embryo they also have enhanced rights because it is about their DNA and their reproduction.

6.7 Molecularization of the Self

Molecularization of human rights leads to another problem, which has been eloquently addressed by Donna Dickenson. If genetic test and self-interest will be the main driving force in health care molecularization, or the promise of the personalized medicine can be easily lead to selfish health care from we to me when people will no longer see healthcare allocated on the human solidarity rather on the actual pre-calculated individual needs based on susceptibility genetic testing (Dickenson 2007). Of course many of these promises are still speculative.

One can argue that even if such a use of genetic knowledge was not foreseen, still general provisions on non-discrimination may be evoked also in this context. If one looks at the wording of the Oviedo convention, it is clear already in Article 1, where purpose and object is specified that the Convention extends its scope of the protection and it does cover also "the dignity and identity of all human beings and guarantee everyone, without discrimination, respect for their integrity and other rights and fundamental freedoms with regard to the application of biology and medicine (Buchanan at al. 2000). While the European Convention on Human Rights speaks on "everyone" which formulation refers to a person, the Oviedo Convention mentions "human being" which provides deliberately a broader and more biological category. This extension of the scope can be seen even more clearly in Article 13 which is about the interventions on the human genome. "An intervention seeking to modify the human genome may only be undertaken for preventive, diagnostic or therapeutic purposes and only if its aim is not to introduce any modification in the genome of any descendants."

One of the most influential and most evident fields in which molecularization and the notion of the human body can be observed and constantly contested is the field of intellectual property (Rose 2007). Under the European Patent Convention[1] even proteins can be protected together with many items of life sciences. In the case law of the European Patent Office one can find several interesting interpretations on what constitutes an invention in life sciences. In the case on Relaxin the Appeal Board had to decide whether the findings of a DNA coding for human relaxin or one of its precursors is a discovery within the meaning of the European Patent Convention. In the T 1836/10 patent case the Appeal Board decided that an intervention may contravene Article 53 (a) and Rule 28(c) EPC whenever a human

[1] European Patent Convention (EPC 2000).

embryo is used as a basic material in a process for preparing embryonic stem cell, even if the embryo is not destroyed. One should note that in this wording although human embryo is considered as a material, still if the use is contrary to "ordre public" or morality the invention is regarded as non-patentable. In other words, in an embryonic level assessment of morality does not depend on the status of person but rather on other specific elements, such as destruction, commercial or industrial use.

6.8 Identification of Sperm Donors

On the most visible consequences of geneticization and molecularization, and it is reflected already in numerous judicial cases, is the issue of gamete donation. If it is, indeed, true that human rights issues are repeated and reinterpreted on the cellular level, it has many consequences. For instance this explains why the legislative processes of the various countries consider egg donation as being different from sperm donation, why surrogacy is forbidden in many places where egg donation is legal and why single women have to fight in a number of countries if they wish to use reproductive services or if they want to have their egg frozen for future reproduction. Since the difference between egg and sperm donation follows from the difference between gender, this distinction, when relevant can be applied also for using human gametes. On the field in which the relevance of identity is increasingly discussed is the anonym gamete donation which a well spread practice in the field of assisted reproduction. The geneticization and molecularization of families can be seen by the growing demands for disclosing the identity of the anonym donor.

In the Canadian *Pratten v. British Columbia* case the main issue was whether the child conceived as a result of *in vitro* fertilization has a right to know the sperm donor. The court compared the differences and similarities with adoption. For instance relinquishing mothers run a significant risk of being stigmatized and this is not the case in sperm donation. Dr. Ehrensaft provided an expert opinion on the relevance of the genetic make-up. "In our Western culture, we privilege the importance of genetic heritage in our family building and in our society at large. Yet an entire group of people conceived with donor gametes anonymously donated are told that for them such roots really do not matter." In Dr. Ehrensaft's opinion, however, the negative psychological effects that may arise from not knowing or not having access to information about one's donor can include identity dysphoria, anxiety, anger, depression, and medical and social angst stemming from being barred access to half of one's genetic history. Such effects can be found in the donor offspring here, including Ms. Pratten. One of the witnesses stated that being denied access to half their genetic history can also create medical risks. In case of gamete donation anonymity and secrecy may create problems. It seems by providing access to identity by those who would like to find out there genetic donor parent it would solve the problems of anxiety.

6.9 Mithocondrium as the 'Second Mother'

Since the Nuffield Council in U.K. has discussed the use and replacement of the mithocondrium in case of some specific inheritable conditions, people refer to mithocondrium as a 'mother'. Personification on the molecular level was obvious even in public debates on the ethics, as artificial fertilization through mitochondrial replacement was called a 'three-parent' procedure involving 'two mothers'. The mitochondrion (organelle) is primarily responsible for the energy supply of cells. Mitochondria have their own DNA indicated as mtDNA. The biological background to the question why the mitochondrion involved in the process is called the "second mother" is that mitochondrial DNA is inherited in a linear fashion, i.e. the genetic data originates either from the mother's or the father's side within a single generation. True, the overwhelming majority of genetic data is stored in the cell nucleus and only 1% in the mitochondrion. Even so, the fertilization procedure that involves the replacement of the mitochondrion has resulted in terms like 'three-parent embryo' or 'three-parent reproduction'.

In the United Kingdom the term 'two mothers' is often used in the media and in public debates on whether to legalize mitochondrial replacement in cases of mitochondrial diseases. In short, the procedure involves the replacement of the mitochondrion of the person diagnosed with the disease with that of another person free from that particular disease. Already in 2012, the Nuffield Council conducted a 6-month public survey about the moral implications of mitochondrial replacement designed to prevent genetic diseases. In August 2012, a total of 1000 people representing various social groups were interviewed at 175 randomly selected locations. The objective of the study was to explore public opinion on the genetic therapy of mitochondrial diseases.

Mitochondrial diseases are severe, genetic diseases that can be cured with mitochondrial replacement. However, a number of ethical counter-arguments, similar to those on cloning, have been presented in opposition to this procedure. One of the strongest counter-arguments is that, due to the mitochondrial replacement, not two but actually three parents are involved in the procedure.

The presence of a foreign human DNA originating from someone other than the parents, however, is not unique. Even though there is the donor's foreign DNA in the blood after marrow transfusion, it is not transmitted to children. With mitochondrial replacement, however, the DNA of the third person is also passed on to children which raises, safety and identity issues, as well.

If, however, we examine these cases from the aspect of mitochondrial diseases that account for a number of infant and child deaths and miscarriages, the guest mitochondrion is more of a therapeutic tool than a peer parent. Lately, it has been proposed that biologists perhaps underestimate the role played by the mitochondria, which, of course, may reshape ethical analysis.

The other counterargument against using donated mitochondria is based on the potentiality of commercialization. True enough, commercialization has been an issue in virtually all reproductive methods. And where an egg from a third person is needed, even the issue of exploitation is raised if someone is forced to "donate" eggs under grave financial coercion (Scully et al. 2010).

I believe that we render human rights, family law and other legal concepts to the human embryo, DNA, egg, blastocyst or mitochondrion without carefully considering the extent to which these mere biological categories differ from colloquial or legal concepts, such as man, mother and child. An extensive consultation process recently completed in 2014 in Great Britain examined whether it was time to regulate mitochondrial donation. This procedure is used in cases of disease inherited via the mitochondria and only in the form of in vitro fertilization. The concept of molecularization shifted the focus from the assessment of health hazards and safety to the identity of the mitochondria. Granting "mother" status to the donor clearly indicates this trend.

6.10 Surrogacy

Surrogacy could have been a balance against the overestimation of genetic ties in the field of assisted reproduction. It could have been an example of a multi-parent model, as surrogacy does involve is birth mother who gives birth on behalf of intending parents, who are usually also genetic parents. Perhaps the legal policy on surrogacy would have been completely different had it not been for the famous Baby M. case of 1987. In the often quoted case, Mary Beth Whitehead was not only a surrogate mother but, since in vitro fertilization was seldom used at the time, also the genetic mother of the baby mentioned only as M. in the court proceedings. After the child was born, lengthy court proceedings ensued between the genetic father and his wife, who were paying for the surrogacy, and the surrogate mother as to who should win custody over the child.

The key moral issue regarding surrogacy is that it turns motherhood into a kind of business enterprise and is coupled with defenselessness, as the biological mother may be deprived of her rights. Surrogacy may be problematic also when the parents "placing the order" and have special wishes or change their minds. Even though the procedure is expensive and the costs are further increased by those of legal services, travel and administration in addition to the fee paid to the surrogate mother and, in special cases, the egg donor, the United States is often chosen as the site for surrogacy due to the high level of organization and the legal environment.

The proliferation of international surrogacy raises another legal issue: if the surrogacy agreement is made abroad, can the child be acknowledged under the law of a country that does not recognize or actually bans surrogacy agreements? Here the issue cannot be settled as simply as with an illegal action under the law of the specific country, as the child would suffer severe legal consequences if he or she could not be granted citizenship in the parents' home country and would have to be smuggled across the border while being sedated with drugs. On the other hand, the notion of surrogacy is hard to reconcile with the legal system in a number of European countries. Although the finding a surrogate mother and making the legal paper work seem to be complicated, it is only after the child's birth that real legal difficulties begin. The father signs a voluntary paternity acknowledgement form and the father's wife usually adopts the newborn child. Even if they have a birth certificate, they are required to have their child recognized by the legal system of their home country after their return from abroad.

In June 2014, the European Court of Human Rights issued rulings in two cases (*Mennesson v. France* and *Labassee v. France*) concerning the recognition of the family status of children born as a result of surrogacy agreements. Upon their return to France both couples encountered difficulties when trying to obtain a birth certificate and citizenship for the child. The local courts and finally the court of appeal all declared the surrogacy agreement null and void as being contrary to French law, under which the carrying surrogate mother is considered the child's mother. The applicants, however, argued before the Strasbourg court that the failure to recognize their children's citizenship was in violation of the New York United Convention on the Rights of the Child and Article 8 of the European Convention for the Protection of Human Rights and Fundamental Freedoms and that it had already been raised in the Genovese v. Malta case that citizenship was an integral part of personal identity. The court ruled that Article 8 was, indeed, violated in both cases and concluded that the child's origin is an important element of privacy rights, which were damaged by the French authorities' failure to recognize the child's relationship with the parents.

With an increase in the number of trans-national cases of surrogacy, the regulations are expected to change, not to mention that surrogacy is, in fact, already permitted in a few European countries. While the combination of genetic and intending parents are recognized as parents the law still cannot provide a consistent answer to how to protect the rights of the gestational mothers, and why embryo's destiny depends largely on the order that initiated their existence. The notion of identity is tied to genetic bonds rather than the element of birth which constitutes another element of molecularization of parenthood.

6.11 Conclusions

As we have seen molecularization has contributed to enormous changes not only in biotechnology but also in the field of legal policy, in social expectations and in even in the extended notion of family. The increasing use of DNA samples in biobanks and in genetic testing have required the elaboration of data protection norms and made it necessary to incorporate new elements into the right to property with relation to human biological samples. The fragmentation and disintegration of the human body makes it necessary to provide different types of control overs specimens, samples, and DNA. As legal policy had to develop often in a hasty manner and as a result did not have adequate time for the sufficient analysis of what and how to protect and to what purposes protect. Legal policies tried to follow the actual science paradigm without much reflection. In addition these legal policies had to balance between two extreme positions, one that claims that human cells, DNA requires the same protection as persons, and the other, which simply ignores the persons behind and takes these samples as mere research tools that were donated for the interests of science and society.

My intention was to challenge both of these positions as in my view my DNA, my samples are not me, they do not possess all aspects of my life, my beliefs, my thoughts, and sufferings. It would be disrespectful to consider persons and biological materials as the same legal entity. On the other side it would be equally unfair and even dangerous to disrobe individual's rights over their DNA. Donation of gametes, cells and DNA should be neither assumed nor trivialized. Individual may retain certain rights or wish to stay in partnership in research and to benefit from research results.

References

Ackerly, Brooke A. 2008. *Universal human rights in a world of difference*. Cambridge: Cambridge University Press.

Agamben, Giorgio. 1998. *Homo sacer*. Stanford: Stanford University Press.

Beyleveld, Deryck, and Roger Brownsword. 2000. My body, my body parts, my property? *Health Care Analysis* 8 (2): 87–99.

Buchanan, Allan, Dan W. Brock, Norman Daniels, and Daniel Wikler. 2000. *From chance to choice*. Cambridge: Cambridge University Press.

Dickenson, Donna. 2007. *Property in the body*. Cambridge: Cambridge University Press.

Duster, Troy. 2003. The genetic screening of 'target' populations. In *Backdoor to eugenics*, 39–59. New York: Routledge.

European Patent Convention. 2000. Convention on the Grant of European Patents (European Patent Convention) of 5 October 1973 as revised by the Act revising Article 63 EPC of 17 December 1991 and the Act revising the EPC of 29 November 2000

Francioni, Francesco. 2007. *Biotechnologies and international human rights*. Oxford: Hart Publishing.

Jasanoff, Sheila. 2011. *Reframing rights. Bioconstitutionalism in the genetic age*. Cambridge, MA: MIT Press.

Joas, Hans. 2013. *The sacredness of the person. A new genealogy of human rights*. Washington, DC: Georgetown University Press.

Laurie, Graeme. 2002. *Genetic privacy*. Cambridge: Cambridge University Press.

Lemke, Thomas. 2011. *Biopolitics: An advanced introduction*. New York: New York University Press.

Locke, John. 1689, 1947. The Second treatise on civil government. In *John Locke: On politics and education*, ed. Howard R. Penniman. New York: D. van Nostrand.

Nussbaum, Martha. 1995. Objectification. *Philosophy and Public Affairs* 24 (4): 249–291.

Rose, Nikolas. 2007. *The politics of life itself: Biomedicine, power, and subjectivity in the twenty-first century*. Princeton: Princeton University Press.

Sándor, Judit. 2012. Bioethics and basic rights: Persons, humans, and the boundaries of life. In *The Oxford handbook of comparative constitutional law*, ed. Michel Rosenfeld and András Sajó, 1142–1165. Oxford: Oxford University Press.

Scully, Jackie Leach, Laurel E. Baldwin-Ragaven, and Petya Fitzpatrick. 2010. *Feminist bioethics*. Baltimore: The Johns Hopkins University Press.

Waldron, Jeremy. 2006. *The right to private property*. Oxford: Clarendon Press.

Weir, Robert F., and Robert S. Olick. 2004. *The stored tissue issue*. Oxford: Oxford University Press.
Zimmer, Franz-Josef, Steven M. Zerman, Jens Hammer, Klara Goldbach, and Bernd Allekotte. 2015. *Protecting and enforcing life science inventions in Europe under EPC and EU law from antibodies to zebrafish*. 2nd, rev and updated edition. C.H. Beck/Hart: München/Oxford.

Legal Cases

Diamond v. Chakrabarty, 447 U.S. 303 (1980).
Mennesson v. France (ECtHR application no. 65192/11, judgment 26.06.2014).
Labassee v. France (ECtHR application no. 65941/11), judgment 26.06.2014).
Moore v. Regents of the University of California, 229 F.3d 831, 84 FEP Cases 129 (9th Cir. 2000).
Pratten v. British Columbia (Attorney General), 2011 BCSC 656 (CanLII).

Part III
Connections: Bioethics and Biopolitics

Chapter 7
Can Boethics Escape from Biopolitics?

Henk ten Have

Abstract The emergence of biopolitics and bioethics are closely related. They not only focus on 'bio,' life in a wide and abstract sense, but also reach beyond the individual subject and beyond theoretical rationality. But instead of developing as two related discourses, bioethics has in fact become the dominant manifestation of biopolitics itself. Bioethics has generated a political regime, a system of governance that administers, fosters and secures life as a form of 'biopower,' controlling the population and disciplining the individual. This emergence of biopower is closely linked to the ideology of neoliberalism since it emphasizes internal regulation by autonomous subjects rather than external force and pressure.

However, this articulation of mainstream bioethics as the prime manifestation of biopolitics, emphasizing humanitarianism, requesting respect for the individual as the subject of politics and ethics, and regulating life from its interior; at the same time neglecting the social, political and economic dimensions of human life, is increasingly more difficult to maintain in light of the globalization of bioethics. Rather than re-iterating and facilitating biopower, global bioethics is better capable of using moral discourse to critically analyze medicine and healthcare. The question is whether global bioethics can escape from biopolitics. Can it develop moral discourses and critical practices that criticize, resist and oppose the dominant neoliberal ideology?

Keywords Bioethics • Biopolitics • Globalization • Global bioethics • Neoliberalism • Patenting • Trade • WIPO • WTO

H. ten Have (✉)
Center for Healthcare Ethics, Duquesne University, Fisher Hall 300, 600 Forbes Avenue, Pittsburgh, PA 15282, USA
e-mail: tenhaveh@duq.edu

P. Kakuk (ed.), *Bioethics and Biopolitics*, Advancing Global Bioethics 8,
DOI 10.1007/978-3-319-66249-7_7

7.1 Introduction

In July 2012, Teresa Stanek Rea, the Deputy Director of the U.S. Patent and Trademark Office (USTPO) testified before U.S. Congress about patents, trade secrets and market access. The chairman of the hearing, Republican Bob Goodlatte started off with stating that foreign governments such as Brazil, Thailand and India block American companies to sell patented products in their countries.[1] Foreign governments undertake unfair activities undermining American innovation and ingenuity. His question was: How can we stop those countries doing that? For more than one hour, mrs Rea directed her ire against the Indian government that had allowed an Indian generic company to produce an affordable generic drug as an alternative to Nexavar, an expensive drug against kidney and liver cancer, fabricated by Bayer AG. She announced that the Obama administration will use all efforts and political pressures to overturn the Indian decision. In a lengthy attack she accused India of violation of WTO rules.[2]

The testimony was hardly noticed in the media since it was overshadowed by the decision of the U.S. Supreme Court to reject some major challenges to the Affordable Care Act (ACA). Obama could proudly declare: "Here in America… no illness or accident should lead to any family's financial ruin." Public and global health experts, however, were appalled. Doctors without Borders called it "really shocking testimony." The information presented is clearly false. There is nothing illegal about compulsory licensing of patented drugs. Countries have every right to do so; compulsory licensing is protected by the WTO. Most shocking perhaps was that there never was any reference in the testimony to cancer patients who needed the drug, or any mentioning of drug prices. Bayer reported $ 4.35 billion net profit in 2013. It is charging $5500 a month for a standard dose of Nexavar, while the Indian generic version costs $173 a month. Rea's testimony was about products and profit, not patients and health.

The hearing also highlighted the hypocritical politics of the Obama Administration. Mark his words: "Here in America." What is good for this country is not the same for other countries. While the Affordable Care Act aims to decrease healthcare costs at home, higher drug prices are imposed abroad. In the same year, the Administration has blocked a proposal in the World Health Assembly to establish a new global research fund to fight disease in the developing world. More or less at the same time, the US government halted any progress in the World Intellectual Property Organization (WIPO) about proposals to lower the price of AIDS medication and

[1] Goodlatte's campaigns were funded by the patent industry. Rea herself is a pharmacist and patent lawyer. In March 2015, Obama appointed a new Director of USTPO, Michelle Lee, a patent lawyer and former Google employee.

[2] For a report on the Rea hearing: http://www.huffingtonpost.com/2012/07/10/obamas-health-policy-global-health-reform_n_1659742.html. For the video of the hearing, see: https://www.youtube.com/watch?v=k9_68z6De9E

other critical drugs in poor countries.[3] It seems that the Obama Administration is captured by the patent industry, especially pharmaceutical companies that do not want any international arrangement that could affect their long-term monopolies. This is a rather consistent policy, despite all rhetoric. A deal was cut with the PhRMA (the Pharmaceutical Research and Manufacturers of America) one of the most powerful lobbies in the US. In return for their support for the ACA, a 12-year monopoly on test data for clinical trials was secured. This will not only prevent the development of new drugs but also raise prices. It violates the Declaration of Helsinki; WTO has never imposed such standards. But in current trade deals, it is now included. The Trans-Pacific Partnership, currently pressured by Obama, has been negotiated in secret; information is not available to the public. Leaked documents show that the issue of intellectual property rights plays a major role. The trade agreement will make the approval process for generic medication more difficult; patents will be longer applicable; knowledge will be restricted. For many countries in the Pacific area it will lead to dramatic increases of life-saving medication. The Transatlantic Trade and Investment Partnership between the EU and the US (TTIP), also negotiated behind closed doors, has the same ingredients. According to Oxfam, TTIP will allow companies to bypass national courts and governments if they permit competition with cheaper versions of medication.[4]

These manoeuvres show a clear policy pattern: Prevent the introduction of generic drugs, impose stricter rules for patents, extend patents, and impose data exclusivity. The rationale is: affordable care for US citizens, not for global citizens. The global standards set are only those of transnational companies. So far, Rea's bravery to overcome resistance ("we will stop this"), has not been successful. In December 2014, the Indian Supreme Court rejected Bayer's appeal against a lower court decision to override the monopoly on Nexavar. It is time that the Indian standards are picked up by other countries like Brazil and South Africa. But where in this story is bioethics?

7.2 Denying Biopolitics

The story illustrates that any debate about the disconnection between biopolitics and bioethics is vain. The lesson apparently is that bioethics consistently denies any connection, and that it hides itself in a discourse that emphatically ignores biopolitical strategies and policies that have any influence on global health that should concern bioethical debate. How can this denial be explained? In my view, there are two explanations. The first is that mainstream bioethics has developed as a specific

[3] For the new global fund, see http://www.huffingtonpost.com/2012/05/25/global-health-fund-obama-administration_n_1544399.html. See also WIPO: http://www.huffingtonpost.com/2012/05/29/trade-deals-cheap-drugs-us_n_1554314.html

[4] For Oxfam's comments on the TTIP, see: https://www.oxfam.org/en/whats-wrong-european-union-united-states-free-trade-talks

concept with a particular theoretical and practical framework. The second is that mainstream bioethics rejects the notion of global bioethics, not recognizing that in the past two decades it has been overtaken by new problems and challenges.

7.3 Mainstream Bioethics

The narrative of the emergence of bioethics is well-known. Medical paternalism, new technologies, and social change in the 1960s and 1970s contributed to undermining traditional medical ethics. Scandals such as Tuskegee called for external regulation, opening up the notion of ethics to a broader range of actors, creating a special discipline involving law, philosophy, theology, social sciences, and humanities. In a surprisingly short time, the new discipline was institutionalized with centres, teaching programs, committees, journals, textbooks and professional organizations. The methodology was canonized by the Belmont Report (National Commission 1979), followed by the subsequent editions of Beauchamp and Childress' classic textbook (first published in 1979). Bioethics was conceptualized as applied ethics, i.e. the application of ethical principles to a variety of practical, biomedical problems. This conception was working with three presuppositions. First, there is a body of principles available to be applied (in fact, varieties of the 'four principles' of Beauchamp and Childress 2013). Second, as a general ethics applied to medicine, the context in which problems arise is not unique. Third, the aim of bioethics is to clarify problems, analyze arguments, weigh alternatives and make recommendations for courses of action, and policies.

This concept of bioethics, as developed within a Western context, and hailed as a typical American invention, was regarded as a model for the entire world. The classic textbook was translated in multiple languages, bioethicists taught courses in many countries, and physicians in Zambia or China recited the principles of autonomy and beneficence. Since the 1990s, the concept became increasingly problematic. It was recognized that mainstream bioethics is a typical reflection of a specific cultural tradition that gives primacy to individual autonomy. The main ethical concern is that decisions concerning health and disease are freely chosen and that harm to others is avoided. The primacy of moral autonomy means that the common good, public interests, community or social responsibility do not figure high on the bioethical agenda. One implication is that the substance of decisions cannot be morally assessed; what matters is whether the decisions are freely made by autonomous individuals. This results in what Daniel Callahan, has criticized as 'minimalist ethics," i.e. a thin ethics that turns ethical challenges into private, not moral choices. Such private morality cannot properly ask questions about the good of society or the proper goals of medical progress, let alone criticize the context within which individual decisions are fostered or impeded (Callahan 2012).

The emphasis in mainstream bioethics on individual autonomy and a thin ethical approach has in fact been an appropriate answer to the challenges that faced medicine and healthcare in the 1960s and 1970s. Never before has medical progress been so impressive. Discoveries and innovations, new technologies and drugs, essentially transformed the practice of medicine. These scientific and technological developments made the traditional ethics inadequate, and called for the new, broader ethical approach. The emergence of bioethics was fuelled by the possibilities as well as limitations of the potentialities of medicine. "Playing God" became a dominant metaphor for the challenges posed to humankind. It is reflected in the milestones often mentioned in the growth of bioethics: the revelation of the Tuskegee study (1972), showing how research can become an end in itself; the Karen Ann Quinlan court decision (1976), showing the need for limiting medical interventions; the birth of Louise Brown by IVF in the UK (1978), showing that 'nature' as limitation is no longer feasible. The source of the ethical problems, therefore, is located in the newly acquired power of science and technology within the lives and wellbeing of human beings who continually are confronted with questions about life and death, disability or normality, and treatment or non-treatment.

Faced with this type of problems, bioethical discourse focused on empowering the individual person, articulating individual rights in opposition to medical paternalism, as well as the significance of rational, voluntary choices among the potential benefits of medical interventions. Mainstream bioethics, therefore, was shaped by its socio-cultural context as well as its problem perception.

7.4 Global Bioethics

Since the 1990s, contexts and problems have changed. Global issues started to emerge, beginning with the HIV/AIDS epidemic in the 1980s. Bioethics also became a global activity with the Human Genome Project and outsourcing of clinical trials. Global activities were initiated by intergovernmental organizations, such as UNESCO and WHO. This expansion was associated with a growing number of cross-cultural studies, outlining the diversity of the global moral landscape. The impact of globalization on bioethics was initially interpreted as a matter of enlarging scope and scale, in other words, mainstream bioethics was globalizing but not fundamentally changing. Over time, this interpretation was more and more difficult to maintain. Scandals and embarrassing cases such as the Trovan case in Nigeria or the placebo trials in Uganda, showed that in global bioethics other issues are at stake: vulnerability, exploitation, injustice, corruption, and double standards. It became more obvious that global problems are not just reformulations of old problems but that they present novel challenges that cannot be met with the usual methods. These observations encouraged the idea that rather than through globalizing mainstream bioethics it is necessary to develop global bioethics as new and broader approach (Ten Have and Gordijn 2013). This is exactly what happened since the turn of the millennium. The number of studies on global bioethics grew exponentially, acrimonious debates scrutinized its strengths and weaknesses, and global bioethical frameworks were declared and promoted. However, most fundamental was the awareness that global bioethical problems differ not merely in scope and scale from the traditional problems of mainstream bioethics but that they are distinct in nature: we are not dealing with the same kind of problems as in the past. While earlier problems have been related to progress in science and technology, global problems are rooted in specific processes of globalization. They are the result of the dominant ideology of neoliberalism (Ten Have 2015). The power of money and commercial interests, rather than the power of medical science and technology produce most of today's bioethical challenges. Focussing on empowering individuals will therefore be inadequate. But even more: as long as mainstream bioethics continues to apply its discourse on global problems, it will operate to facilitate the further expansion of this neoliberal ideology.

7.5 Scrutinizing the Context of Bioethical Problems

Basically, this was the message of Van Rensselaer Potter (1988) when he re-coined bioethics as 'global bioethics.' What is necessary, as he argued, is not a reformulated medical ethics, but a really new and more encompassing approach that could address the global problems of poverty, pollution, inequality and injustice, war and violence. It should focus on the social, economic and political context in which these problems arise. It should also include the environmental problems exacerbated through globalization. When Potter introduced the notion of global bioethics, major environmental disasters shocked the world. Rachel Carson has published *Silent Spring* (1962), explaining the negative and long-term effects of chemicals on the environment. The oil tanker Torrey Canyon polluted the British coast (1967); in Seveso, Italy, thousands of people and animals were contaminated with dioxin (1976); and in Bhopal, India, approximately 20,000 people died due to toxic gases (1984). Bioethics, therefore, can no longer focus on relations between individuals but has to attend to social and environmental conditions in which health and healthcare are seriously affected.

Potter's appeal to a novel approach is currently more and more recognized. In principle, globalization has the potential to benefit everybody. In practice, however, it is associated with rising vulnerability, inequality, marginalization, and exclusion (Sassen 2014). Global mobility and interdependency are asymmetrical. Healthcare workers trained in developing countries migrate to more developed ones. Wealthy patients travel to poor countries for advanced private care while the majority of the population in those countries have less access to public care. Body parts and organs flow in the other direction. New medication is tested on vulnerable populations who will never have a chance to use the drugs themselves.

The source of today's global bioethical problems is not simply globalization but specific processes of globalization which are primarily driven by economic motivations. The world is first of all regarded as a global market. The 'market' has become the dominant metaphor for the organization of social life. Everything, from transportation and research to healthcare and education should be transformed into commodities and services transacted in a market (Lafontaine 2014). Competition rather than cooperation is the core value. In this perspective, markets should be 'liberalized,' i.e. all constraints and regulations should be removed or reduced, so that individual consumers will be able to make the most efficient choices.

The negative impact of neoliberal policies on global health is now undeniable (Davies 2010; Homedes and Ugalde 2005). Social inequalities have increased. In the US (2012), the top 1% claimed 23% of national income (twice as much as in 1980). At the same time, real wages have stayed put or fallen. Poverty has increased; it has the 4th highest rate of poverty among OECD countries; over 30% of child poverty; after Chile, it is the most unequal country in the world. The social fabric responsible for equality of opportunity is gone (Putnam 2015). The same argument is used in Europe; solidarity which used to be a principle of pride is dead. Growing inequality is now considered as a major risk for political stability, undermining democracy and political institutions (Schui 2014). The damaging effects of neoliberal policies are particularly clear in the breakdown of healthcare systems in many developing countries. Reduced expenditures for health and social services, privatization of care, lower salaries for healthcare workers, and introduction of user fees for patients have reduced access to health services for the majority of populations. In Tanzania, for example, life expectancy decreased from 50.1 years in 1992 to 43.1 years in 2002. In Latin America, health reforms increased inequity and inefficiency. Entire populations are deprived from necessary treatment and medication, simply because the prices are unaffordable. Globally, each year, 2 to 3 million people die of tuberculosis; they could be treated but 79% of them do not have access to appropriate medication.

7.6 Global Bioethics

Against this backdrop, global bioethics is redefined as a *social* ethics. If global problems are produced within a context of neoliberal policies and practices, it will be inevitable to critically address this context (Brody 2009). Unless, of course, bioethics chooses to persevere in its bio-political strategy to focus on scientific and technological issues emerging within such context, and ameliorate the way individuals seek to cope with them. Effectively, this is the bio-political agenda of mainstream bioethics: leave the social, economic and political determinants of health outside of its scope, because the assumption is that they are either unalterable or require political, not ethical action. Bioethics operates as common instrument of a global moral economy (Salter and Salter 2007).

Global bioethics has developed, and is further developing, two modes for a different approach of global problems: a broader moral discourse, and an enlarged repertoire of practices.

7.7 Broader Moral Discourse

Although global bioethics incorporates the value of personal autonomy and individual rights, it has introduced a moral discourse with a more extensive horizon (Ten Have 2015). First, it argues that human beings are not abstract and de-contextualized individuals; they are necessary embedded in social structures. Social, cultural, economic and political conditions make flourishing in health possible. Second, it emphasizes that living together means that human beings share heritage and common goods. Common interests are not the aggregation of private interests. Individual persons as citizens are concerned with the common good; they are not merely rational consumers interested in maximizing benefits for themselves. Third, though individual actions and concerns are important, they will not bring about social transformation. Neoliberal globalization is not a fact of nature but the result of deliberate decisions of policy-makers, often without any involvement of populations. New forms of collective engagement and agency will be necessary to influence the systemic conditions producing global problems. This is precisely what the logic of neoliberalism tries to prevent, arguing that individuals have moral responsibility to care for health and the environment, and promoting that they make 'ethical' choices (organic and labelled food; generic drugs or 'ethical' clothing).

Critique of neoliberal globalization has re-activated a broader set of moral concepts such as human vulnerability, social responsibility, sharing of benefits, and future generations. These concepts will direct the bioethical attention to structural determinants of health rather than individual decisions concerning care. Other concerns can therefore be on the agenda; moral issues can be framed in a different way; other solutions and approaches can be proposed.

7.8 Enlarged Repertoire of Practices

Global bioethics not only is a theoretical discourse but also includes a broader variety of practices as usual in mainstream bioethics. One example is advocacy. Many NGOs used this as an instrument to press for changes not in rhetoric but in concrete practices, for instance in pressuring governments to translate the right to health in better access to medication (Ho 2013). Advocacy sometimes requires language and actions that go beyond what is traditionally acceptable in the academic discourse of mainstream bioethics. Jennifer Chan (2015, p. 177), in her book on AIDS activism highlights the international trade system, and the role of WTO, as "the rot at the core of global governance today." Blaming economic injustice and structural violence as basic problems is unusual in bioethics but it opens up new ways to analyze and redress global problems. Another is capacity-building. Its significance is recognized in mainstream bioethics but it is usually regarded as 'empowerment': the capacities of individuals as responsible agents of change should be improved. The assumption is that improvements will only result from enhanced individual efforts. Global bioethics argues hat capacity-building should be based on promoting the common good, and will require collective action. It is primarily a matter of global injustice and structural inequalities that requires a '*moral* structural adjustment,' such as fair benefit sharing with populations and communities aiming at strengthening a health-care system that would benefit the entire populace.

7.9 Global Bioethics and Biopolitics

How does global bioethics relate to biopolitics? If it blames mainstream bioethics for being biopolitics, how can it escape biopolitics itself? With neoliberalism as the dominant political ideology and as the source of many global bioethical problems, how can global bioethics be not political? Neoliberal globalization is the result of a political project since the 1980s; how can it effectively be criticized and perhaps transformed without bio-political strategies?

The answers to these questions obviously are negative. Global bioethics is at the same instant necessarily biopolitics (Bishop and Jotterand 2006). It does not attempt to separate ethics and politics, mimicking the efforts of mainstream bioethics and at risk of losing its normative discourse. It recognizes that ethics and politics are intimately connected; normative considerations are central to both. Its basic concerns are not technical issues but political ones.

First of all, global health is a political concern (Davies 2010). Though manifested in individuals, health is determined by social, cultural and economic factors that require political and collective action. Health can also be framed in different ways. As a security problem for example it focuses on threats and protection, rather than human needs. Infectious diseases such as Ebola should then be contained by controlling travel and migration; there is no urgency to address the root causes. Securitization of health therefore introduces the language of friends and enemies; it concentrates on infectious diseases as the main problem of global health.

Secondly, global governance is a political rather then technical challenge. Although it often takes a technocratic approach, proceeding from 'facts', engaging 'experts', and establishing 'committees,' this type of administrative governance is inadequate at the global level, as the mismanagement of the 2014–2015 Ebola epidemic demonstrates. Mainstream bioethics assists in focusing this type of governance on the challenges and hopes of science and technology, in connection to the individual decision makers, diverting attention from the social and political implications of biotechnology and medical science, and certainly from the economic context. Recognizing that the broader setting of governance is dominated by neoliberal discourse, will allow identifying processes that have been outside of the scope of bioethics. Most of all, it will allow analyzing these processes as normative influences that have been taken for granted. The recent interest in 'bioeconomy' is an example. In the OECD worldview, human beings as living organisms are themselves sources of renewable energy. This is not a merely biological perspective (viewing living beings as coping with entropy, and capable of regeneration and self-repair) but also an individualizing one with strict normative implications. If each body has bio-value, each person should act as an entrepreneur who must invest in his or her biological capital. Responsible bio-citizenship means taking care for the body and bodily health, not improving social conditions. Governance implies that the subjective life typical for human beings (bios) is separated from life as the vital force of the human species (zoe). Articulating individual responsibility effectively evades any concerns about the political context in which trade has priority over health. This context, however, cannot be avoided. In the global perspective, issues such as inequality, exclusion, exploitation, and vulnerability continue to arise, directing attention to the normative settings of the context itself.

Third, global responses such as foreign aid and humanitarian assistance are political rather than technical issues. Aid can be misused and diverted to the benefit of one party. Aid can prolong conflicts. Assistance may be limited by governments. The traditional humanitarian principles of impartiality, neutrality, and independence are compromised today. It is not possible to remain neutral when confronted with atrocities. There is a need to speak out and testify of suffering (Terry 2002). Humanitarian action therefore is political; it requires going beyond the perspective of the individual sufferer. Continuing to focus on emergency relief and saving lives, without addressing the social, economic, and political conditions that produce vulnerabilities and structural injustices, is in fact a political choice. Biological life rather than political life with its history of violence is prioritized (Fassin 2009).

Fourth, many bioethical problems are identified and addressed through 'globalization from below,' i.e. through grassroots organizations and local NGOs. Global principles are used to challenge local practices while local knowledge feeds back into the global framework. However, the context within which global and local ethical approaches interact is thoroughly political (Lipschutz 2005). Civil society and particularly NGOs, are often regarded as guardians of the public interest, as a third sector between state and market. Now that neoliberalism has privatized the public sphere and the impact of the state on society is minimized, NGOs are helpful because they can step in to provide services that otherwise are not profitable. But this role is more palliative than therapeutic. The gaps created by neoliberal policies are filled in; public services are first reduced, and now compensated by social movements. More and more NGOs no longer criticize the underlying power structures (Choudry and Kapoor 2010). They emphasize ethics as personal commitment and individual lifestyle, promoting the idea that small-scale, ethical initiatives can bring social transformation (growing your own food or riding your bike against climate change). Social ills are reduced to personal morality (Sharzer 2011). Furthermore, many NGOs have turned towards technical and managerial approaches rather than critical examining systemic conditions of power and inequality. They are increasingly cooperating with and co-opted by sponsors that significantly contribute to causes for which the NGOs were initially established. Finally, new forms of partnership in global governance have faded the difference between corporations motivated by profit and social movements representing vulnerable populations. Not all NGOS are the same: there are distinctions between governmental NGOs, corporate NGOs and people-oriented NGOs. The power relations between these different bodies and organizations are not equal, even when they are partnering.

As these examples illustrate, global bioethics operates in a context that is political and with concerns that are political.

7.10 Dealing with Biopolitics

Biopolitics, in the description of Thomas Lemke (2011) is "a new political knowledge and new discipline… to "govern" individuals and collectives." It is a new form of power, working not through external but internal regulation; it is social control through self-conduct and self-care. This form of power is associated with the emergence of neoliberalism. Bioethical discourse is helpful since it teaches individuals to subject themselves to neoliberal values. The rational, autonomous individual is the entrepreneur of himself (Lemm and Vatter 2014; De Beistegui, Bianco and Gracieuse 2015).

These interpretations of biopolitics do not sufficiently take into account the contemporary global context. It presents this context as homogeneous and universal, assuming that many processes of globalization are driven 'from above,' by 'global players such as states, international organizations, transnational companies, and NGOs. In reality, there are different levels of political activity. Politics is not a matter of states. Many activities in global bioethics are undertaken 'from below.' The global dissemination of human rights for example is due to localized social movements, specific political struggles and resistance against humiliation and oppression. As Slavoj Žižek (2011) argues, a political act transforms the singular into the universal. An excluded population does not claim so much its own rights as well universal rights for everyone. The universality of the claim transforms their claim into a political one. Offering a global discourse with universal claims provides global bioethics with a radical opening: the possibility to create a new political order. It is using Dewey's concept of 'social intelligence." (Dewey 2000). Intelligence is not a private property. It is manifested in science and research: collective efforts to produce reliable results. Social intelligence is the expression of interdependence and participation, articulating not consensus but compromise. It broadens the horizons so that participants will have a different view of problems and possibilities of resolution. This should be reflected in democracy: not simply counting votes but participation in communal life with a focus on the common good. Social structures are not unchangeable.

7.11 Conclusion

Global bioethics can be a critical discourse. The condition is that it continues to introduce new concepts and language, going beyond the narrow settings of academic discourse, and mobilize a broader repertoire of activities, recognizes the need for negotiation, and collective responsibility and action, and scrutinizes the distribution of power at a global scale. It should first of all unravel the ideological notion that the market is neutral, showing that it is always favouring the interests of some.

References

Beauchamp, Tom L., and James F. Childress. 2013. *Principles of Biomedical Ethics*. New York: Oxford University Press. (7th edition) (first edition published in 1979).

Bishop, Jeffrey P., and Fabrice Jotterand. 2006. Bioethics as biopolitics. *Journal of Medicine and Philosophy* 31: 205–212.

Brody, Howard. 2009. *Future of bioethics*. New York/London: Oxford University Press.

Callahan, Daniel. 2012. *The roots of bioethics. Health, progress, technology, death*. Oxford: Oxford University Press.

Chan, Jennifer. 2015. *Politics in the corridor of dying AIDS activism and global health governance*. Baltimore: Johns Hopkins University Press.

Choudry, Aziz, and Dip Kapoor, eds. 2010. *Learning from the ground up. Global perspectives on social movements and knowledge production*. New York: Palgrave Macmillan.

Davies, Sara E. 2010. *Global politics of health*. Cambridge/Malden: Polity Press.

De Beistegui, Miguel, Giuseppe Bianco, and Marjorie Gracieuse, eds. 2015. *The care of life. Transdisciplinary perspectives in bioethics and biopolitics*. London/New York: Rowman & Littlefield.

Dewey, John. 2000. *Liberalism and Social Action*. Amherst: Prometheus Books. (original 1935).

Fassin, Didier. 2009. Another politics of life is possible. *Theory, Culture & Society* 26 (5): 44–60.

Ho, Cynthia M. 2013. Beyond patents. Global challenges to affordable medicine. In *The globalization of health care. Legal and ethical issues*, ed. I. Glenn Cohen, 302–317. Oxford/New York: Oxford University Press.

Homedes, Nuria, and Antonio Ugalde. 2005. Why neoliberal health reforms have failed in Latin America. *Health Policy* 71: 83–96.

Lafontaine, Celine. 2014. *Le corps-marche. La marchandisation de la vie humaine à l'ère de la bioéconomie*. Paris: Seuil.

Lemke, Thomas. 2011. *Biopolitics. An advanced introduction*. New York/London: New York University Press.

Lemm, Vanessa, and Miguel Vatter, eds. 2014. *The government of life. Foucault, biopolitics, and neoliberalism*. Bronx: Fordham University Press.

Lipschutz, Ronnie D. 2005. Power, politics and global civil society. *Millennium – Journal of International Studies* 33 (3): 747–769.

National Commission for the Protection of Human Subjects of Biomedical and Behavioral Research. 1979. The Belmont Report: Ethical principles and guidelines for the protection of human subjects of research. *Federal Register* 44 (76): 23191–23197.

OECD. *The bioeconomy to 2030. Designing a policy agenda*. OECD: Paris, 2009 (http://www.oecd.org/futures/long-termtechnologicalsocietalchallenges/42837897.pdf).

Potter, V.R. 1988. *Global bioethics. Building on the Leopold legacy*. East Lansing: Michigan State University Press.

Putnam, Robert D. 2015. *Our kids: The American dream in crisis*. New York: Simon & Schuster.
Salter, Brian, and Charlotte Salter. 2007. Bioethics and the global moral economy. The cultural politics of human embryonic stem cell science. *Science, Technology & Human Values* 32 (5): 554–581.
Sassen, Saskia. 2014. *Expulsions. Brutality and complexity in the global economy*. Cambridge/London: The Belknap Press of Harvard University Press.
Schui, Florian. 2014. *Austerity; the great failure*. New Haven: Yale University Press.
Sharzer, Greg. 2011. *No Local. Why small-scale alternatives won't change the world*. Winchester/Washington: Zero Books.
Ten Have, Henk. 2015. Respect for human vulnerability: The emergence of a new principle in bioethics. *Journal of Bioethical Inquiry* 12 (3): 395–408. doi:10.1007/s11673-015-9641-9.
Ten Have, Henk, and Bert Gordijn. 2013. *Handbook of Global Bioethics*. Vol. 4. Dordrecht: Springer Publishers.
Terry, Fiona. 2002. *Condemned to repeat? The paradox of humanitarian action*. Ithaca/London: Cornell University Press.
Žižek, Slavoj. 2011. *Intolerantie*. Amsterdam: Boom. (original: 'Multiculturalism, or, the cultural logic of multinational capitalism,' *New Left Review* I/225, 1997: 28–51).

Chapter 8
Bioethics as Politics

Tuija Takala

Abstract Bioethics has never been far removed from politics or political agendas. For instance, many of the early contributors argued fiercely for the rights of women, or the rights of unborn babies, depending on which side of the political debate their thinking fell. The empowerment of patients against the prevailing medical paternalism was similarly politically motivated. What separated the early bioethical contributions from the purely political ones was that they were using established theories and methodologies of their own disciplines to argue for their case. The scientific validity of these claims was something that could be assessed against the theoretical background of the respective disciplines. Bioethics has always been an inter- and multidisciplinary enterprise and this tendency has become more and more prevalent. The ever-changing realities of medicine and biomedical sciences, available resources and the views, values and beliefs of the various stakeholders are playing an increasingly important role in bioethical deliberations. Widening and deepening the understanding of the complex issues by utilizing a wide variety of approaches is, of course, laudable. However, when the aim is not only to describe, but to prescribe, the multidisciplinarity becomes a methodological problem. Politics can be defined as the practice and theory of influencing people and that is, I would argue, what most bioethicists are trying to do. They have a moral stance and related policy recommendations that they are trying to convince people to accept. This tendency is further strengthened by the expectation from the funding bodies for research to yield practically relevant results. The chapter substantiates the claim that (much of) bioethics is politics by looking at the kind of work that is being carried out in bioethics, the speculative nature of the cost-benefit analyses at the heart of bioethics debates and the problems caused by inter- and multidisciplinary approaches with an emphasis on new technologies.

Keywords Bioethics • New technologies • Interdisciplinarity • Policy

---

T. Takala (✉)
School of Business, Department of Management Studies, Philosophy of Management,
Aalto University, P.O. Box 21230, FIN-00076, Aalto, Finland
e-mail: tuija-maija.takala@aalto.fi

P. Kakuk (ed.), *Bioethics and Biopolitics*, Advancing Global Bioethics 8,
DOI 10.1007/978-3-319-66249-7_8

8.1 Bioethics Literature

Matti Häyry and I recently edited two special issues for the journal *Bioethics*: *Best Practice in Conceptual Philosophical Bioethics*[1] and *The Role of Philosophy and Philosophers in Bioethics.*[2] During the editorial process we browsed through all the c. 1000 articles published in the journal since 1987 and only about 50 could be considered theoretical, or not primarily practice oriented. A couple of years earlier when I was trying to identify methods of philosophical bioethics, a literary survey of the *Kennedy Institute of Ethics Journal* (1991–2011), conducted by Johanna Ahola-Launonen, produced similar results.

Already a quarter of a century ago, Ronald M. Green recognized the "policy- and legislative-orientation" as one of the main reasons for "theoretical weaknesses in bioethical theory and methodology".[3] I would tend to agree with him that bioethics done with practice in mind "often shows a willingness to set aside conceptual issues in the name of a consensus policy".[4] Whilst practical solutions might indeed be what the world needs, compromising on the theoretical, methodological and conceptual levels means that bioethics loses its scientific justification and comes closer to politics.

8.2 Ethical Analyses in Bioethics

Whatever the underlying theoretical framework, most bioethical analyses contain a balancing exercise of benefits and harms. Obviously this is most prominent in consequentialist studies, but it is present in other approaches as well. In consequentialist decision making there are four key elements: (expected) benefits, dampening factors, financial costs and (risks of) harm. The expected benefits are the reason why a particular course of action is being suggested. When we are talking about the new biotechnologies, these can include beneficial advances in medicine, pharmacy, agriculture, the food industry and the preservation of our natural environment. Normally the expected benefits boil down to increased human wellbeing (either directly, as in terms of better treatments, diagnostic methods and nourishments, or indirectly, as in overcoming pollution and preserving the environment.) The dampening factors refer to those prevailing rules, practices and arrangements that are independent of the considered course of action, but tend to counteract its benefits either by lowering their quantity or quality, or by promoting their unequal distribution. Profit seeking and the existing local and global inequalities are among the issues to be considered.

[1] http://onlinelibrary.wiley.com/journal/10.1111/%28ISSN%291467-8519/homepage/best_practice_in_conceptual_philosophical_bioethics.htm

[2] http://onlinelibrary.wiley.com/journal/10.1111/%28ISSN%291467-8519/homepage/the_role_of_philosophy_and_philosophers_in_bioethics.htm

[3] Green (1990).

[4] Ibid.

Any new technology is expensive and the financial costs and their legitimacy need to be considered. The risks of new technologies can be felt on individual, local, national and global levels. The containment and release (intentional and unintentional) of genetically and synthetically modified molecules, cells and organisms can have unexpected and uncontrollable adverse consequences. Additionally, there are possible untoward social, political and economic ramifications – some of these caused by the aforementioned dampening factors – that need to be considered.[5]

There is then the further question of the acceptability of the risk and its acceptability to whom? In analyzing risks we have to consider their probability, magnitude and tolerability. Understanding probabilities are difficult[6] and interpreting them even harder. An individual is far more likely to get killed in a car accident than to win the lottery or be involved in a plane crash. Yet few of us are afraid of being in traffic, but many are afraid of flying and even a greater number of people keep playing the lottery. When it comes to the magnitude of possible harm and the new biotechnologies, making predictions is equally problematic. Theoretically, introducing new molecules, cells or organisms into the existing biota could have catastrophic consequences. We have seen this happening with unmodified plants and animals when they have been introduced to new biosystems. Further, living organisms evolve through mutations and it is very difficult to predict the direction these take and what the consequences turn out to be. Then again, the odds of a global disaster are not huge and there are ways of minimizing the risk of such an event, and even if something harmful starts to spread, we probably would not be unarmed in fighting it.

The tolerability of the risk is partly connected to balancing the risks of harms with the expected benefits. Similarly to risks, the probability and magnitude of the benefits are difficult to predict and interpret. In addition, the distribution of the possible benefits will depend on a number of factors beyond the new technologies. The unequal distribution of wealth, locally and globally, the disparate health care systems, the different political establishments and other institutional factors will very likely contribute to an unequal distribution of both harms and benefits. The tolerability of a risk has much to do with the one bearing the risk seeing it as a worthwhile risk. That is, do the potential benefits outweigh the risks in the considered view of the "risk taker"? And given that the risks of modern biotechnologies could well touch all of us, a democratic decision-making procedure is called for.[7] The general public needs to become more aware of the new technologies to be able to meaningfully contribute to the choices made.[8]

[5] Häyry and Takala (1998).

[6] Gigerenzer and Edwards (2003).

[7] Obviously democratic decision-making has all sorts of problems, the least of these not being very able to protect minority views.

[8] See e.g. Playing God: The Rock Opera, an attempt to reach people and get them engaged in bioethical deliberations. playinggodrocks.com

With both the expected benefits and the risks being uncertain, the balancing remains speculative and will be strongly influenced by one's beliefs concerning the possibilities of science and one's view of the human nature. Much of it has to do with whether one is an optimist or a pessimist. Similar facts will represent themselves differently depending on whether one has a positive or a negative view of the future.[9] Many people who advocate the new technologies speak as if the benefits are certain and the risks negligible, while those with a more pessimistic view tend to focus on the uneven distribution of benefits, if any, and on the risks.

8.3 Argumentative Strategies

The most vocal contributions to the ethical debates tend to be either strongly against or adamantly for the new technologies. This tendency is furthered by the popular media, which prefers headline material.

The precautionary principle is often evoked when the goal is to criticize the new technologies. The principle, very roughly, states that if an action A could have catastrophic consequences we should refrain from doing A even if the realization of those consequences is uncertain.[10] As most modern technologies could, theoretically, have catastrophic consequences to human health and the environment, it is not difficult to see how this argument appeals to those who are already skeptical about the technologies. On the other hand, if one happens to have a more positive view of the future, a contrasting argument, the hopeful principle, can be brought into play. The hopeful principle states that if something very beneficial could follow from B we should do B even if we cannot be totally certain that the benefits will follow.[11] The rationale here is that if, for instance, genetics could give us a cure for cancer, genetically modified crops could feed the hungry and synthetic organisms could eat away pollution, these would easily outweigh any associated costs or risks and therefore we should continue with these endeavors.

Obviously these are both predominantly rhetoric tools in which certain aspects are exaggerated and others overlooked. Pro-technology arguments tend to overstate the likelihood and proximity of the benefits, dismiss the related justice concerns and downplay the risks, while the critics often oversell the risks, highlight the potential injustices and overlook the possible benefits.

The slippery slope argument has often been used in end-of-life discussion and when arguing against genetic engineering. The empirical version of the argument

---

[9] Häyry and Takala (1999).

[10] Häyry (2005). A more detailed formulation would be: "When our actions could be harmful, but this harmfulness cannot be verified or falsified by scientific inquiry, the burden of proof is, according to the principle, on those who propose such actions. Until further research shows that the actions do *not* have the suspected ill effects, they should be disallowed. Häyry (2010).

[11] Holm and Takala (2007).

goes roughly as follows: While introducing practice A would be morally acceptable, A would, as a matter of empirical fact lead to a morally unacceptable B, and therefore we should refrain from A also.[12] Or as explained in another context by Häyry:

> People who are suspicious of the technological change tend to emphasise the bad effects of the new developments. They admit that gene treatments can have desirable aims, but they remain concerned about 'slippery slopes', which would lead from good, intended results to bad, unintended ones. Curing diseases like X-SCID would in and by itself be desirable, but allowing gene therapies for them could gradually lead to other, more dubious genetic interventions and eventually to plainly unacceptable practices like germ-line enhancements. Moral values, according to the advocates of this view, are more important than physical and mental wellbeing, and in any assessment of harms they should be taken fully into account.[13]

Much has been written on the problems of the slippery slope argument and the two main lines of attack are: either questioning the causal connection between the acceptable and the unacceptable, or challenging the immorality of what is deemed unacceptable.

Lately, those promoting the new technologies have started to use an argument, which, in terms of structure, closely resembles the slippery slope argument. According to the "automatic escalator argument" there might be moral costs associated with pursuing technology C (say, stem cell research), but since C would, as an empirical fact, lead to D (say, cure for cancer), which is a great moral good, we should move forward with C regardless of the costs.[14] Again, as with the slippery slope argument, the causal connection can be disputed. Further, it is unclear whether the proposed route to D is the only possible one and thereby, the possibility of other, less morally costly, methods is overlooked. Also, while a cure for cancer would arguably be a truly great good, there are other goods which might be even greater and further, the distribution and availability of D might make it a lesser good than it is made to seem.[15]

On the critical side, both the precautionary principle and the slippery slope argument are sometimes made stronger by introducing the argument from "Playing God".[16] The religious versions of this argument have some weight within theological debates and they resonate with religious people – You should not play God! – but as philosophical arguments, they are weak. The secular version uses the omnipotent and omniscient "god" as a metaphor. The idea is that when you do not know all and are unable to control everything, you should not do things that could have unwanted and unforeseeable long-term consequences. This version of the argument has some more substance to it, but in the end, it comes very close to the precautionary principle and is open to similar criticisms. The notion of naturalness is also frequently evoked.[17] We have a very deeply rooted intuition about there being a link between

---

[12] See note 11, p. 3.

[13] Häyry (2010), p. 193.

[14] Ibid.

[15] Ibid.

[16] Chadwick (1982) and Häyry (2011).

[17] The ideas of playing god and naturalness have also been seen in the literature to form a partnership: "The charges of 'acting unnaturally' or 'playing god' evoke lofty images of environmental

what is unnatural and what is immoral. However, as any even preliminary analysis will show, the connection between the two is shaky to say the least. On the one hand, there is much unnatural in what we consider to be good, such as most Western medicine, and on the other, we would hardly deem "natural selection" as moral.[18] There is no straightforward link between unnatural and immoral, however, regardless, appealing to what is natural is a powerful tool when one wishes to sway people.

While the critics use arguments from playing god and naturalness, the champions of the new technologies appeal to people's benevolence. Let me call this the "Bad Samaritan argument". The claim here is that those impeding the development of the new technologies are actually harming any number of people who could be helped, cured or fed, if the technologies were allowed to be developed.[19] This accusation oversees the possibility that the technologies might not reap the benefits they promise and it gives no room for other ways of benefitting people. For instance, if the goal is to reduce suffering, the funds used for developing the technologies could arguably provide for more instant and tangible results if used in more conventional ways. This would also make the distribution of benefits more transparent. Much of what is argued for in bioethics comes down to the speaker having a way with words. Bioethicists market their own moral and political beliefs by seemingly clever arguments.

8.4 The Problem of Interdisciplinarity

What has blurred the line of academic bioethics and political bioethics further has been the emergence of interdisciplinary bioethics. At its best, interdisciplinarity allows us to draw from the methodological and theoretical strengths of several disciplines to bring together a more substantial body of knowledge and understanding than we could ever have reached by utilizing the tools of a single discipline alone. However, when unrelated elements are hand-picked from several disciplines, the theoretical and methodological foundations on which the overall results are based become debatable. The results will always depend on the underlying assumptions and on the sources chosen. And since there is no established methodology for making these choices, the scientific reliability of the results remains challenged.[20]

---

disaster and divine punishment. If we change the course of rivers or move mountains, the long-term cumulative consequences can be catastrophic in ways which we cannot precidely define in the light of our current knowledge. And if we choose to create 'abominations' like animal-human hybrids, clones or parentless children, not only are the material outcomes unpredictable, but we can also have to confront the wrath of gods." Häyry (2010, p. 132).

[18] Takala (2004).

[19] Some go even further and argue, not only should we allow the technologies to be developed, but that we should actively aid in making this happen by becoming research subjects. Harris (2005).

[20] Takala and Häyry (2014).

In practice, it seems to me, that the most common method in interdisciplinary bioethics is to start with a normative conclusion and then to gather evidence from a number of sources to justify one's chosen normative views. Obviously this works well for commissioned studies where the direction of the policy recommendation is pre-set. Social scientists can study the tolerability of the proposed policy and try to identify the least alterations or other causes of action needed to the policy that make it acceptable to people. Legal scholars can study the legal boundary conditions and philosophers can complete the job by providing the normative justifications.[21] However, bioethics with a political agenda is first and foremost politics. The problem with such studies being considered as academic is twofold. First, without an established methodology, it is difficult to say whether the conclusions are well supported or not, so assessing the scientific validity of interdisciplinary contributions is difficult. Secondly, after they have been accepted as a part of bioethics literature, they tend to contribute to the field expanding it rather than advancing it. The same bioethical arguments are presented over and over again in slightly different contexts and building on previous work becomes difficult. There is no established theoretical or methodological framework for interdisciplinary bioethics. And in the absence of that, bioethics is more like politics.[22]

8.5 The Rightful Scope of Normative Claims

Purely theoretical, conceptual or descriptive bioethics aside, everything in bioethics is political. At its core, bioethics is about rights and responsibilities, justice and entitlement – all of which are political notions. Any bioethicist trained in philosophy has the tools to defend or criticize any bioethical position from a number of angles.

However, when it comes to our basic beliefs – whether we are liberal or conservative, religious or secular, and so on – the choice has very little to do with our academic training. Our basic normative beliefs are no better or worse than anyone else's, but when we speak as experts in ethics, we tend to forget that. People who are highly educated are often better equipped to argue for their normative positions, but when it comes to the most profound ethical beliefs, there is no reason to presume that those of an academic would be more reliable than anyone else's. Just look at the variety of academic positions on the new technologies from Jürgen Habermas'[23] very critical views to John Harris',[24] "let's do everything we can and more" attitudes. There is no overarching "scientific" argument to tell which of these views should triumph.

[21] Takala (2015).

[22] For a philosopher's critical take on this, see e.g. Häyry (2015).

[23] Habermas (2003). See also, Fukuyama (2002); Kass (2002) and Sandel (2007).

[24] Harris (2007). See also, Agar (2004); de Grey and Rae (2007) and Glover (2006).

Academic training in bioethics gives us many tools for expanding our knowledge on the complex nature of bioethical dilemmas. We can provide new arguments, viewpoints, considerations and insights, but when we start giving unconditional, or categorical normative answers, as we like to do, we step away from our field of expertise and become part of the politics of bioethics.[25]

Acknowledgement This chapter was produced as a part of two Academy of Finland projects, Methods in Philosophical Bioethics (SA 131030, 2009–2014) and Synthetic Biology and Ethics (SA 272467, 2013–2017), and of the Finnish Cultural Foundation Argumenta project Justice and Its Alternatives in a Globalizing World. The author acknowledges the Academy's, and the Cultural Foundation's, support with gratitude.

References

Agar, N. 2004. *Liberal eugenics: In defence of human enhancement*. Oxford: Blackwell.

Chadwick, R. 1982. Cloning. *Philosophy* 57: 201–209.

de Grey, A., and M. Rae. 2007. *Ending aging: The rejuvenation breakthroughs that could reverse human aging in our lifetime*. New York: St Martin's Press.

Fukuyama, F. 2002. *Our posthuman future: Consequences of the biotechnology revolution*. London: Profile Books.

Gigerenzer, G., and A. Edwards. 2003. Simple tools for understanding risks: From innumeracy to insights. *British Medical Journal* 327: 741–744.

Glover, J. 2006. *Choosing children: The ethical dilemmas of genetic intervention*. Oxford: Clarendon Press.

Green, R.M. 1990. Method in bioethics: A troubled assessment. *The Journal of Medicine and Philosophy* 15: 179–197. 194.

Habermas, J.. 2003. *The Future of Human Nature*. Trans. W. Rehg, M. Pensky, and H. Beister. Cambridge: Polity Press.

Harris, J. 2005. Scientific research is a moral duty. *Journal of Medical Ethics* 31: 242–248.

———. 2007. *Enhancing evolution: The ethical case for making better people*. Princeton/Oxford: Princeton University Press.

Häyry, M. 2005. Precaution and solidarity. *Cambridge Quarterly of Healthcare Ethics* 14: 199–206.

———. 2010. *Rationality and the genetic challenge: Making people better?* 191. Cambridge: Cambridge University Press.

———. 2011. *Playing god: Essays on bioethics*, 47–54. Helsinki: Helsinki University Press.

———. 2015. What do you think of philosophical bioethics? *Cambridge Quarterly of Healthcare Ethics* 24: 139–148.

Häyry, M., and T. Takala. 1998. Genetic engineering and the risk of harm. *Medicine Health Care and Philosophy* 1: 61–64.

———. 1999. Biotechnology and the environment: From moral objections to ethical analyses. In *Ethics: Proceedings of the 20th world congress of philosophy*, ed. K. Brinkmann, 169–178. Bowling Green: Philosophy Documentation Center.

Holm, S., and T. Takala. 2007. High hopes and automatic escalators: A critique of some new arguments in bioethics. *Journal of Medical Ethics* 33: 1–4.

Kass, L.R. 2002. *Life, liberty, and the defense of dignity: The challenge for bioethics*. San Francisco: Encounter Books.

[25] Takala (2005).

Sandel, M. 2007. *The case against perfection: Ethics in the age of genetic engineering*. Cambridge/London: The Belknap Press of Harvard University Press.
Takala, T. 2004. The (im)morality of (un)naturalness. *Cambridge Quarterly of Healthcare Ethics* 13: 15–19.
———. 2005. Demagogues, firefighters and window-dressers: Who are we and what should we be? *Cambridge Quarterly of Healthcare Ethics* 14: 385–388.
———. 2015. Get to the point! Philosophical bioethics and the struggle to remain relevant. *Cambridge Quarterly of Healthcare Ethics* 24: 149–153.
Takala, T., and M. Häyry. 2014. Neuroethics and animals, methods and philosophy. *Cambridge Quarterly of Healthcare Ethics* 23: 182–187.

Chapter 9
On the Relationship Between Bioethics and Biopolitics: What Bioethics Can Learn from Biopolitics

Darryl Gunson

Abstract This paper discusses the conceptual and practical relationship between bioethics and biopolitics. The paper starts with reference to Lemke's claim that bioethics (as a field, compared to biopolitics) narrowed the discussions on life and politics to a discussion on values. Then, it proceeds by identifying certain ideal-types about the nature of bioethics which, it is argued, illustrate two ends of the spectrum upon which discussions of the field of bioethics occupy. They are constructed according to two criteria: the first concerns the scope of bioethics, which includes the aims of bioethics and its problems; the second involves the details of the method(s) of bioethics including the use of empirical data. With respect to the scope of bioethics, two models are discussed which offer a broader and a narrower focus for bioethics. The first suggests that bioethics is, or should be, a global undertaking, with a broader and renewed focus on global justice, basic health needs, and our relation to the environment. The second suggests that bioethics is primarily concerned with a 'northern' agenda, which focuses on the problems with healthcare systems and issues thrown up by advanced biotechnology. With respect to methods, the idea of bioethics as a purely, or primarily, philosophical enterprise is discussed and a more interdisciplinary model is sketched. These versions of bioethics are used as the basis for a comparison with biopolitics.

The paper continues by examining some prominent (although by no means all) characterisations of biopolitics and compares them with the bioethical models. 'Biopolitics' is used to refer to many issues that range from bioterrorism and security issues to biotechnological developments and the ethical issues that arise from these, and from the nature of state control and surveillance over its citizens, to a concern with the rise of social philosophies such as neo-liberalism. One possible interpretation of the relationship between bioethics and biopolitics is that they cover much the same ground, but that somehow biopolitics is broader and deeper, focusing on the wider political context of policy and regulation, and that this is not part

D. Gunson (✉)
School of Media, Culture and Society, University of the West of Scotland,
Paisley PA5 2BE, UK
e-mail: darryl.gunson@uws.ac.uk

P. Kakuk (ed.), *Bioethics and Biopolitics*, Advancing Global Bioethics 8,
DOI 10.1007/978-3-319-66249-7_9

of the remit of bioethics. However, it is argued in this paper that this has some plausibility only if we adopt an unnecessarily narrow definition of bioethics in the first place. Once we shift our focus in bioethics to appreciate the necessity for broadening our horizons – perhaps to a 'global bioethics' – and once we understand the necessity for empirical research from all relevant sciences, we begin to see that bioethics does have a legitimate concern with the type of questions and analyses hitherto associated with the field of biopolitics.

Keywords Neoliberalism • Global bioethics • Lemke • Critical biopolitics

9.1 Introduction

Whether bioethics is regarded as an academic discipline, or as a more loosely connected area of scholarly activities, it is only relatively recently that the term has become common and recognised in universities and other institutions. The term probably dates back to the late 1960s or early 1970s when, according to one commentator, 'bioethics' was first used in discussions relating to the formation of the Kennedy Institute of Ethics at Georgetown University.[1] The mission of this institute, and the influence it has had on other similar institutions, has shaped the way that bioethics has been practiced, by defining the kinds of subject matter appropriate for investigation and by influencing the self-understanding of those who see themselves as bioethicists. At around the same time, Michel Foucault[2] introduced the similar term 'biopolitics' into academic discourse, and whilst he was not the first to use the word – it had already been used to describe various approaches to life and politics – it has since then come to be understood to refer to a range of social science approaches that focus on various aspects of the political struggles surrounding life generally, and the loss of control that people may experience over their lives due to state and other influences. The focus of this emerging field ranges from concern for the environment to the forms of self-surveillance by which modern people routinely monitor their behaviour, and it also includes the effects on self-understanding and identity when whole populations become the objects of study. Thomas Lemke writes:

> In the past ten years intellectuals inside and outside the United States have used the notion of biopolitics to reflect on issues as heterogeneous as the war on terror after 9/11, the rise of neoliberalism, and biomedical and biotechnological innovations such as stem cell research,

[1] Jonsen (1998) claims that the name 'bioethics' was mentioned in conversation between Dr. Andre Hellegers and representatives from the Joseph P. Kennedy Jr. Foundation – Mr. R. Sargent Shriver and his wife Eunice Kennedy Shriver. The conversation was about sponsorship of Hellegers' proposed institute which was to be devoted to the study of religious and ethical aspects of medical and biological science. The term subsequently became associated with the Kennedy Foundation in the early 1970s and its mission to lead the way in what many considered a new field of enquiry.

[2] Foucault et al. (2008).

> and the human genome project. In these debates, the concept of biopolitics has often served as an interpretive key to analyse how the production and protection of life is articulated with the proliferation of death; or it seeks to grasp how the reduction of human beings to "bare life" (e.g., in Guantánamo and Abu Ghraib) is linked to strategies to optimize and enhance human capabilities and life expectancy (Lemke 2011, xi)

Furthermore, the concept has also been invoked to discuss political asylum policies, as well as the prevention of AIDS with questions as diverse as financial support for agricultural products, promotion of medical research, legal regulations on abortion, and advance directives of patients specifying their preferences concerning life-prolonging measures, also falling under the broad understanding of what is biopolitical.

Bioethics exhibits a similar diversity of meaning and application that seems to overlap with the brief sketch of biopolitical issues. Ranging from an initial concern almost exclusively with the ethical problems associated with new medical and biotechnological advances, bioethics now increasingly encompasses broader public and global health issues, as well as concerns with 'dual-use technologies'. Given this, one might expect that bioethics and biopolitics would have much in common and that their respective endeavours may even be complimentary.

On the face of it, the fact that bioethics and biopolitics both contain the word 'bios' might suggest that they are in fact studying the same thing – life – perhaps from slightly different perspectives: one considering the ethical implications of, say, developments in the life sciences and the other focussing more on the political context of translating the results of ethical discussions into actual policy. It would therefore be tempting to suggest that the relationship between the two fields is one of overlapping and mutually complementary emphasis, perhaps focussing on different aspects of a common project, with an appropriate division of labour. This thought would be reinforced by a cursory glance at the list of applications of the term where both bioethicists and those who see themselves as doing biopolitical work would recognise topics and concerns listed in the 'other camp' as also being their own.

However, such a cosy division of labour does not really capture the relationship between the two enterprises, or at least as it is understood by recent commentators, and there remains a suspicion that bioethics is somehow narrower and less reflexive than its biopolitical counterpart and, rather than being complementary to the concerns of biopolitics, actually serves to distort discourse about human life. Lemke expresses this reservation about bioethics thus:

> Bioethics has narrowed the terms of the debate on the relation between life and politics since discussion is mainly conducted in ethical terms as an argument about values. (Lemke 2011, 123)

In this chapter I address this claim of Lemke's (and its broader implications) and consider what the relationship between bioethics and biopolitics is, or should be. Is biopolitics part of a broad bioethical field with its own contribution to a common project, or is the relationship the opposite with bioethics subsumed under the boarder biopolitical enterprise?

9.2 Bioethics: Retrospective

In order to understand the relationship between bioethics and biopolitics it is necessary to describe the respective fields in more detail. I do this by identifying the things that are typically studied, the methods used, and any normative orientation, implicit or explicit, that seems to prevail. I begin with bioethics.

Bioethics, as noted, is a relatively new field, probably not much more than 40 years old. One influential account of the history and development of the field is Albert R. Jonsen's *The Birth of Bioethics*. This work presents the history, subject matter and methodology of bioethics from a perspective that is heavily influenced by the development of bioethics in the United States of America. According to Jonsen, 'bioethics' was first coined in the early 1970s, just before the Kennedy Institute was set up at Georgetown University. From this point on bioethicists pursued an agenda that was dominated largely by developments in medical science and cognate areas, and the ethical problems that such advancements raised. Typical topics for consideration were: the ethics of research into human subjects; the ethics of genetics; eugenics; genetic engineering; and mapping the human genome. The ethics of organ transplantation and artificial organs, the supply of organs and the ethical issues around selection of patients for a scarce resource are also core topics. The ethics of death and dying and the problems associated with defining and determining death are also core, including issues arising from our increasing abilities to resuscitate patients, and, more generally, advances in intensive care. Of course, another related and definitive topic is the ethics of allowing death: euthanasia and assisted suicide. At the opposite end of the spectrum work has focussed on the ethics of human artificial reproductive technologies. Bioethics was largely about ethical problems associated with technological advancement in the life sciences.

The methods for addressing these questions were seen, at least initially, as largely philosophical, imported from moral philosophy and applied ethics. Where the literature addressed methodology at all, it tended to be dominated by discussions of Utilitarian and Deontological approaches as well as Natural Law reasoning, the 'New' Casuistry, and, of course, the 'principlist approach' developed by Tom Beauchamp and James Childress (1979/2012). This is not to say that all was harmonious; the field has continued to exemplify an acute sensitivity to almost all of those initial defining precepts, as could have been expected given that the style of inquiry was modelled on that of moral philosophy and applied ethics. Nonetheless, it is not too much of an exaggeration to say that the principlist approach and its concomitant idea of a 'common morality' dominated the field, at least in the early years. Even when the discussions of individual topics were not confined to the application of the 'Georgetown mantra' – autonomy, beneficence, non-maleficence and social justice – this tended to set the terms of the debates. As such, it is understandable that some might concur with Lemke's view of bioethical inquiry being largely focussed around the content and application of ethical values and principles.

The view of bioethics presented by Jonsen, and implicit in Lemke's view, is largely retrospective, looking back at the way in which bioethics developed, particularly in the US. Insofar as it captures what is true of the early development of the field, particularly in the US, it is no wonder that much of the theoretical discussion did centre on values and principles and on the philosophical debates regarding the merits, status and meaning of these. As such, there is some truth in Lemke's assessment. However, whatever the institutional origins of 'bioethics' are, and despite the predominantly philosophical focus, particularly in the early years, the field has expanded and broadened its horizons significantly. The idea that bioethics is mainly about values is, at best, outdated and fails to appreciate how the field has developed over recent years.

9.3 Bioethics: Looking to the Future

If we allow that Jonsen's book captures much that defined the bioethical field as it developed, we should also note that another influential work takes a rather different stance and argues for a vision of bioethics that looks to the future, rather than reporting on the past trajectory as a US-dominated field. Van Rensselaer Potter's book *Bioethics: looking to the future* presents a view of bioethics that shifts its focus away from the 'northern agenda' of ethical problems posed by new biotechnology towards a broader vision that has as its central themes the very survival of the human species (Potter 1971: 1). The expanded vision advocates a field that is ready to harness 'biological values' and scientific knowledge to address a much wider range of issues that have implications for the future safety and survival of the species. Of central concern here is the environment in the broadest sense, not least because it sustains life and there are now well-known threats to its ability to do this. The new vision of bioethics also stresses the use of 'dangerous knowledge' and how it is important to understand the significance of new knowledge and its potential misapplication; many technologies have 'dual-use' (1971: 69).

Another issue of global concern given prominence in the new agenda is population control. Potter suggests that the world population is out of control and, with advances in medicine that prolong life and enhance our ability to reproduce, it could be catastrophic without the measures of birth control. The new vision therefore emphasises the need to move away from a preoccupation with the technologically driven questions that are of interest primarily to wealthy states and argues for a renewed engagement with the priority problems of our time: population; peace; pollution; poverty.

Given this expanded agenda and the call for a refocussing of the objectives, it is also clear that, methodologically speaking bioethics could not remain a purely ethical enterprise. It has had to embrace a broader range of approaches to these issues. That is not to say that there is no ethical dimension to these new problems. There will continue to be ethical work to be done whether that be in articulating the ethical frameworks at work in the global community or assessing the acceptability of solutions and policies. What it does mean is that solutions to bioethical problems cannot be discussed solely in terms of the applicability of values and principles, and this is something that the bioethical community are aware of and which has been a long standing part of the reflexive self-understanding of the discipline.

9.4 A More Expansive Notion of Bioethics

Jonsen and Potter present different views of what bioethics has been and what it should be, and even though there might be a certain tension between the two accounts, both perspectives do capture something important about bioethics and continue to have a significant influence on bioethics, as it is practiced and written about. The tension between focussing on 'high tech' medicine and a sensibility that seeks to refocus towards broader, perhaps global, issues has itself been fruitful and productive. Even when bioethics was going through its 'philosophical phase' in its early development, and whilst the 'principlist approach' was relatively fresh and influential, it has at the same time demonstrated a remarkable capacity to reflect on the assumptions and arguments that seemed to be shaping the character of the field. For example, there was much debate about the principles – not only what they mean and how they are to be applied, but also whether there are other principles that could be added to the list.[3]

That debate led to suggestions that there may in fact be regional forms of bioethics based on variations in principles and values that constitute the 'common morality' of the regions.[4] However, whether or not there really are regional moralities underpinned by values operating in a region or associated with distinct cultures, as in the claim that there are 'Asian values',[5] one upshot of this is that bioethicists are now more sensitive to the dangers and problems associated with applying values and principles uncritically, which has further contributed to the expansion and refocussing of bioethics.

This shift has been a steady movement away from the narrow focus prominent in Jonsen's account, rather than a dramatic change of priorities. For some, however, the change in bioethical priorities is not quick enough with Chattopadhyay and De Vries (2008) complaining that the 'colourful cakes' for wealthy consumers – pre-implantation genetic diagnosis, neural enhancement, genetic therapy and possibly enhancement – detract from more basic global, yet no less ethical issues such as poverty, health inequality, basic health care, AIDS. Indeed, in a similar vein, Hellsten (2008) also points to the division of bioethical interest between the West with its fascination with issues such as abortion, euthanasia and access to(expensive) reproductive technologies, whereas elsewhere in poorer regions and the developing world access to basic healthcare and to treatments for curable diseases is more of a priority.

---

[3] See for example: Häyry (2003).

[4] See for example: Rendtorff and Kemp (2000).

[5] Wulff (1994) and Kam-por (2002).

Of particular importance here is the role that international organisations such as UNESCO have come to play in shaping the scope and application of bioethical work.[6] With its various universal declarations, such as the Universal Declaration on Bioethics and Human Rights, we see attempts to address Potter's[7] recognition that our survival depends upon a common value system. UDBHR aims at framing a set of universally applicable principles around which consensus could be built. With its harnessing of bioethical topics to the discourse on human rights, we see an attempt to create a consensus not only around global principles and values, but also around the recognition of what the important substantive topics are.[8] Insofar as this has been successful we see an important step away from the parochial preoccupation with the 'northern agenda' and the focus on values and principles mentioned by Lemke, to a more global approach to dealing with the problems of the age which is more reminiscent of Potter's expanded refocused agenda for bioethics. Values and principles are still important, but the bioethical discourse is not dominated by them.

9.5 What Should We Do?

Despite the ongoing transformation of bioethics, it is clear that the central questions remain strongly normative. Bioethics may have expanded its boundaries significantly so that, alongside questions such as the ethical status of embryos and the moral acceptability of things such as preimplantation genetic screening and diagnosis, there is a growing concern with global issues such as organ trafficking, HIV, pandemics, research ethics, access to basic health care, which are also currently priority concerns for the World Health Organisation (WHO).[9] The questions that some of these issues raise may seem to be better answered by technical experts, with very little discussion required regarding values or principles. This is true up to a point, but even questions regarding how best to halt the spread of HIV cannot be adequately addressed outside the context of values and norms that inform behaviour. Furthermore, sometimes values and norms may need to be subject to critique, as solutions to problems may come into conflict with local ethical values. The point is that even when the course of action may seem obvious and clear, and the normative direction relatively uncontentious, this is not quite the same as merely uncovering facts about the situation. And sometimes, of course, the normative issues are large and contentious such as those associated with organ trafficking, where the big normative issue is whether or not it is ethical at all. The idea that bioethics is not *just* about values and principles, should not be understood as implying that bioethics no longer needs to engage in such discussions.

---

[6] UNESCO (2006).

[7] Potter (1971: 192).

[8] Andorno (2007).

[9] See http://www.who.int/ethics/topics/en/ accessed 08.06.2015.

The ethical background to all bioethical work is a concern with what we should do: Should we do X? What should we do about Y? Ought we to allow Z? Despite the shift in the priorities and central issues of bioethics, it still retains its ethical, normative orientation. In so far as one tries to answer such questions it is possible to do so in a more, or less, hypothetical manner. The hypothetical approach would structure questions conditionally without really needing to know what the actual facts are. Its logic would be something like the following: if the facts are thus, and the prevailing values are so, then it is permissible to do z; or, if the consequences of doing x are y, and the consequences of doing a are b, then on balance it would be better to do y. This is valuable work as oftentimes the facts are simply not known, or the relevant values and ethical considerations unclear. But, even then, some appreciation of the possibility/probabilities is desirable if not strictly necessary.

However, with such normative questions the force of any conclusions that have a practical orientation will be increased if we know such things as: what the consequences of doing X actually will be; Does it go against deeply held principles? Is it really possible? What are the social/political implications of doing X? In order to answer these kinds of questions we need the methods and input from many forms of inquiry: philosophy, science (medicine and others), social sciences (including sociological and political inquiry). I take no stand here on the exact division of labour involved, but whatever the answer to the question of the relationship between facts and values, I take it that the more relevant empirical information one has about a topic, the better the normative conclusions will be. Indeed, it is now widely accepted that bioethics cannot be pursued as a purely ethical enterprise and that bioethics is moving out of its purely hypothetical phase and increasingly engaging with knowledge from other disciplines.[10]

One such area that bioethics ought to be sensitive to concerns the political dimensions to problems and their solutions. It is tempting to think that one should draw a distinction between questions that are purely ethical (in principle) and questions to do with the practicality or political difficulties of a course of action or a solution to a problem, and to say that bioethics is only concerned with the former and not the latter. However, enticing such a temptation may be, it should be resisted. It is through engaging with the sociological and political dimensions of a topic that we can come to understand what actions, policies, solutions are actually possible. If normative commitments are constrained by empirical possibilities, then facts about how people behave that may be gleaned from sociology, or information about the political dimension of a problem, are important to bioethical discussion. If 'ought' implies 'can' then rich data about what is possible will be vital for bioethical inquiry if the conclusions it reaches are to be relevant and useful in achieving good ends.

From this brief survey, coupled with the suggestion that, contra to what is implied in the quotation from Lemke, bioethics is not only concerned with values, we can see that, at least on the face of it, bioethics has a legitimate interest in the politics of life. This is because the ethical question of 'what should we do' cannot be answered fully without knowing more about the context and circumstances that give rise to

[10] See, for example: Levitt (2003), Hedgecoe (2004), and Levitt (2004).

bioethical questions, and this is something that a hypothetical discussion of values cannot provide by itself.

9.6 Biopolitics

Thomas Lemke (2011: 3) provides a useful initial taxonomy of views and practices that may be called biopolitical. He distinguishes between what he calls naturalist conceptions and politicist orientations. The former takes life as the *basis* of politics, and the latter takes life as the *object* of political processes. The naturalist tradition ranges from early 'organicist' theories of the state,[11] the ideas in the twentieth century about race which were exemplified in national socialism, as well as in other movements, to modern, largely discredited, 'biologistic' ideas in political science which suggest that behaviour is much less free than is usually assumed and is, to a large extent, biologically determined.[12] Insofar as we may wish to follow Lemke here, we should note that this aspect of the biopolitical is definitely something that is also part of the bioethical enterprise as well, because at the heart of the construction and critique of ethical arguments, is the questioning of hidden assumptions in arguments and, especially, discredited theories. Thus, it is fair to say that bioethics does have a legitimate interest in this aspect of biopolitical history and theory and the critique thereof, especially if these views form part of, or are assumed by, ethical arguments.[13]

By 'politicist' Lemke means the view that biopolitics is a sub-branch of politics more generally, that is largely concerned with the regulation of different aspects of life. He further identifies two different ways in which this approach has been pursued: ecological and productivist. The ecological approach is largely defensive, seeking to recruit politics as a discipline and as action to the end of protecting or preserving the natural environment. This view is perhaps in line with what most people would understand by the terms 'environmentalism' and 'ecology', and seems to have some similarity with the expanded vision of Potter, and the broadening of the bioethical agenda, insofar as it is concerned with the natural environment and the implications that environmental harm may have for human (and perhaps animal) well-being.

The productivist approach, on the other hand, includes the views of those who are less focussed on preservation and more open to 'dynamic expansion and productivist expansion'. Details aside, this last concern is very closely attuned to the new

[11] See page 9 for a discussion of the ideas of the Swedish Political Scientist Rudolf Kjellen.

[12] See Chap. 1 *passim* for an extensive discussion, where the point is made that the biologistic tendency in political science assumed a false biological determinism as its corrective to what it perceived as an excessive focus on 'free agents'.

[13] The interest that bioethicists have, quite rightly, shown in contemporary discussions regarding the implications of epigenetics are an example of bioethics having a legitimate interest in ideas that Lemke identifies as part of the history of what has been understood as biopolitics.

biotechnological developments in the medical and life sciences, and Lemke further contends that the development of scientific knowledge, especially genetics, has resulted in a blurring of the boundary between nature and society, and that the bedrock of nature that was characteristic of the conservative ecological biopolitics is now less certain and clear cut. The emphasis is now on the generative possibilities of the new technologies and on how what was once fixed and 'natural' is now, potentially, malleable according to human design. As Lemke himself argues:

> As a result of this problematicisation, the ecological version of biopolitics was weakened until ultimately it was integrated into the technocentric variant. If the former assigned itself a task that tended toward the conservative and the defensive, pursuing the goal of preserving natural foundations of life, the latter is more dynamic and productivist, concerned with the exploitation of these foundations. (28)

The first two chapters of Lemke's book provide a rich and detailed account of the complex variants of these themes that he has identified. The striking thing about this account is that the questions and issues that so define biopolitics here also exhibit a striking similarity to the kind of thing that is part of main-stream bioethical discourse. In as much as bioethics has come to recognise that, along with Potter, the environment is of central ethical concern, and that we should not be too hasty in upsetting natural equilibria, it overlaps significantly with the ecological variant of biopolitics. To the extent that bioethics is still concerned with the social, ethical and legal impact of new technologies such as genetics, it clearly encompasses much that is definitive of the productivist approach to biopolitics as well. Lemke himself admits as much:

> Such questions led ultimately to a second stratum of meaning in biopolitics, one that is situated close to the considerations and concerns of bioethics. These relate to the collective negotiation of, and social acceptability of what is technologically possible. (26)

With respect to the original question regarding the relationship between bioethics and biopolitics, we see that taking into account the fact that bioethics itself has moved away from its early preoccupation with ethical values, even though, quite rightly, it has not abandoned this concern, and has begun to discuss a much broader range of issues, we can see that there is significant overlap between the ethical and the political aspects of 'bios'. Furthermore, bearing in mind the normative orientation of bioethics – just because we can, ought we? – we can also see that the biopolitical is part of the bioethical enterprise. Insight and understanding regarding the political dimension to these concerns is, ultimately, grist to the mill of the ethical enterprise of answering the questions about what we ought to do.

9.7 Critical Biopolitics

However, despite the considerable overlap between the two fields and the possible division of labour based on the desirability for more information to render the ethical arguments more practically compelling, there persists the view that biopolitics is somehow more fundamental than bioethics. Lemke identifies a tradition of 'critical' biopolitics writers who seem to offer something much deeper and nuanced than the previous naturalistic and politicist readings and which may be the source of the view that biopolitics is concerned with issues that are even more fundamental, in some sense, than the normative ethical concerns of bioethics.

Critical biopolitics is exemplified in the writings of witers such as Foucault et al. (2008);[14] Agemben (1998); Hardt and Negri (2000); Rose (2007). Certainly these authours are concerned with life and politics, but they are also all quite different from each other as well. What they do seem to have in common, according to Lemke, is that each, in their own way, recognises that biopolitical phenomena are a product of social processes, rather than immutable laws or other constants, whether those be social or natural. Moreover, it is one of the critical functions of biopolitical work to address the contingency of such processes and the phenomena they give rise to and to show how they incorporate institutional preferences and normative choices. In other words, the deficiency of non-critical bioethics is precisely that it takes for granted something that the new generation of thinkers are keen to deconstruct – the categories of life and politics themselves. Concepts of 'life', 'nature' even 'political' are not isolated concepts but historically interdependent – 'relational' – and the new critical bioethics aims to account for the instability and fragility of this relationship, and how life and politics may be historically related.[15]

One might try and illustrate this concern in the following way. What is considered 'natural' or 'normal' may be offered as a kind of normative base-line, something that has to be preserved or even achieved. This kind of reasoning is widespread in bioethics with, for example, objections to technological developments such as human reproductive cloning sometimes proceeding on the basis that cloning is not natural. Objections to certain forms of human enhancement have also invoked similar arguments such that enhancements are not natural and the subsequent people may not be normal.[16] But what is 'normal' and 'natural' is, not stable and is the result of political processes and the deployment of scientific knowledge. It is well-known now that one of the things that Foucault himself identified as definitive of biopolitics was the rise of the social sciences and the shift from the individual to the population as the object of knowledge; measuring population characteristics can lead us to see what is normal and is a tool for reinforcing behaviours, which in turn can shape how people understand themselves and their lives.[17]

Here one might ask whether or not this is something that bioethics is, or should be interested in? Is this the deep nuanced insight that biopolitics reveals, to which

---

[14] Naturally one would point towards the early texts such as Foucault et al. (2008). But most of Foucault's oeuvre could be considered as biopolitical in the 'critical' sense.

[15] See Lemke (2011) generally and especially the introduction and Chaps. 3 and 9.

[16] Indeed, invoking notions of naturalness or normality are not only used to resist new possibilities, but also may be used to justify new practice on the grounds that it is merely an extension of what is normal or natural.

[17] This idea may also be described in the language of 'social constructivism' where categories, for example gender, are constructed rather than given and immutable.

bioethics is blind? Perhaps it is, but it is worth noting that the debate about human enhancement which has been prominent in bioethics journals over recent years is but one example of bioethical inquiry which demonstrates a significant appreciation of the contingency of the natural and the political. This is particularly notable in the discussions about the social, ethical and legal implications of altering what has hitherto been thought of as natural – the human genome. Bioethicists are attuned to the prospect of human choice, whether individually or collectively, affecting what is thought of as natural or normal through the realisation that it is within our grasp to alter human beings at the genetic level. Another example of bioethical willingness to entertain the contingency and constructed nature of some of our basic social categories is the debate over cloning, where the question of whether it matters if a clone is not the biological offspring of the donor, but of the donor's parents, has been discussed.

9.8 Foucault: Radical Sociology of Knowledge and Subjectivity

Through his lectures and later writings Foucault has been one of the most influential thinkers associated with critical biopolitical theory. Biopolitics for Foucault is associated with a particularly modern form of power, where power and knowledge are inextricably bound together, such that knowledge itself is shaped and determined by power relations which, in turn, perhaps simultaneously, are themselves shaped and developed by the uses to which such knowledge is put. There are at least three separate but ultimately related phenomena that Foucault refers to as biopolitical: first, the historical shift in the form that power typically takes in society. In the West this is characterized as the movement from a society based on the absolute repressive power of the sovereign, to a more subtle form of 'biopower' where the state, using the fruits of the emerging social and medical sciences, is more able to intervene and regulate life. A second form focuses on the role of biopolitical mechanisms in the emergence of racism in the modern period, and a third refers to a distinctive form of governing populations associated particularly with liberal forms of self and social regulation.

Arguably, what these three themes share is a reliance on a cluster of basic ideas. First, as already mentioned, knowledge and power are interlinked. Second, the new disciplines of the modern period, including statistics, epidemiology, demography, biology, psychology, are implicated in the exercise of power through the 'government' of *populations,* which is achieved through various scientifically informed processes of categorization, correction, exclusion, normalization and discipline. It is precisely the utility of the new sciences for discovering the 'nature' of populations that underpins the possibility of shaping population behavior and also self-regulation. Perhaps one of the most important insights here is the idea that knowledge of populations can shape individual and group self-understanding, particularly when found in conjunction with processes of normalization and the like.[18]

Of course this is a very brief sketch, but it provides just enough detail to see that Foucault's insights, as important as they are, are not necessarily completely antagonistic to the bioethical project. To take an example, consider what we might call the rise of 'autonomy' and of the 'autonomous individual', understood as the particular form of self-understanding, increasingly prevalent in modern societies, whereby collectivities and wider social structures are thought to be less important in determining the trajectories of individual lives than the decisions and choices that autonomous agents make. No doubt this appears, to some extent, entirely natural and normal; people are individuals with their own desires, preferences and life plans and they should be allowed, with minimal interference from others, to pursue those. This view has become so entrenched, both as an aspect of the way people understand themselves, but also as a basic ethical principle, such that it is reflected in what many regard as the most important principle of bioethics: the principle of autonomy, which is at the heart of the principlist approach, as well as being a key component of liberal political philosophy.[19] The Foucauldian approach would seek to understand how the 'autonomous individual' came into being in the first place, and this would no doubt reveal the complexity of the social processes and contingent events that contributed to people thinking of themselves and others in this way.

There are two points to be made here. First, the historical narrative as to how the 'autonomous individual' came to be conceived may well demonstrate that this was not a necessary progression of the idea of freedom, or some other teleological notion, but is the product of historical contingency, and power. This may also make us more sensitive to alternative forms of self-understanding. However, none of this undermines the bioethical project, focusing as it does on practical questions. The detailed knowledge that such a historical understanding yields finds its place in the bioethical approach as part of the project's commitment to scrutinizing the dominant ethical values and their role in society. But, for all that, the practical questions that bioethicists usually have to address must *start* with norms and values (and identities) as they are, at present, rather than how they might be.

The second point is that such an historical approach might well serve as a corrective to a bioethics that takes for granted prevailing norms and values, and as such it could be the spur to a greater and deeper reflexivity. Nonetheless, as valuable in general as such correctives might be, it is worth noting that bioethics as currently practiced is already highly reflexive in this sense. Even though the practical questions of bioethics often start with a consideration of the relevant existing values and principles, this has not prevented bioethicists from engaging in critique of those values. The idea of autonomy as the dominant ethical principle of our day, for example, has received more than its fair share of criticism and scrutiny.

---

[18] See Foucault (1980: 144) and Lemke Ch 3: 39.

[19] For a discussion see: Gillon (2003).

9.9 Lemke: Biopolitics as 'Bioanalytics'

These Foucauldian ideas underpin Lemke's own preferred account of biopolitics, what he calls 'bioanalytics'.[20] This is meant to be a kind of synthesis of previous forms of biopolitical work, which he differentiates from the politicist and naturalist conceptions as focussing on, "…neither the causes nor the effects of the politics of life, describing instead its mode of functioning." (118) In its concern to describe how the politics of life functions bioanalytics has as one of its central tasks the job of showing how the regulation of life processes is bound up with social discourses and how these may give ride to new forms of identity and self-understanding. This aspect of the biopolitical and its place in an expanded bioethics has been touched on above, but Lemke gives further substance to this central idea, advising us that bioanalytics should account for the phenomenon of subjectivisation. That is:

> The manner in which subjects are brought to work on themselves, guided by scientific, medical, moral, religious and other authorities and on the basis of socially accepted arrangements of bodies and sexes…how are people called on, in the name of (individual and collective) life and health (one's own health and that of the family, nation, "race" and so forth) to act in a certain way…how do subjects modify scientific interpretations of life for their own conduct and conceive of themselves as organisms regulated by genes, as neurobiological machines, as composed bodies whose organic parts are, in principle, exchangeable? How can this process be viewed as active appropriation and *not* as passive acceptance? (120)

Lemke also argues that critical bioanalytics must identify and understand what he calls biopolitical 'regimes of truth' which are characterised by the answers to questions such as: Who are the experts and who has authority? What knowledge do they have? What language – vocabulary is used? What problems – definitions are accepted as valid? Coupled with this, bioanalytics must also contend with 'strategies of power' and how these relate to, amongst other things: structures of inequality; what forms of life are regarded as socially valuable; how domination, exclusion, racism, and sexism are inscribed in the body; who profits and who bears the costs of processes of regulation and physical improvement. These tasks clearly reflect the central Foucauldian ideas that: knowledge and power do not exist, nor can they be understood, in isolation from each other, and that these are also intimately linked to various ways that we come to understand ourselves and others. Put like this, biopolitics might seem to offer a perspective that really is at odds with bioethics. At its heart is a commitment to seeing the basic categories of any discourse as constructs that need to be understood, and possibly deconstructed. One might say that bioethics relies for its normative arguments on categories that are relatively stable, whether these be existing values and principles, or even dominant conceptions of self and identity found in the public at large, whereas biopolitics is always on the verge of destabilising these categories, digging deeper than its bioethical counterpart.

[20] See Chap. 9, (119 *passim*).

9.10 Power and Subjectivity

Bioethics can and should dig deeper than merely pointing out which practical conclusions are implied by which accepted norms. But, how much deeper should it dig? If it is to provide rationally defensible answers to the practical questions it encounters, then bioethicists do at least need to acknowledge that some norms and values may need to be abandoned, modified, or supplemented with others. Fortunately, questioning prevailing norms, values and rules is a core component of mainstream bioethics. Furthermore, if the question is: 'what should we do?', or 'should we do x?' then anything that is relevant to answering the question rationally is part of the enterprise. This means that bioethics does need to be a truly interdisciplinary endeavour. And this includes taking note of the radical critical insights of biopolitics.

Such insights may play a role in assessing the consequences of proposals, not just for individuals, but at the level of populations – both physically and socially. Likewise, the exercise of power is also of vital important. For example, how governments or elites (e.g. Big Pharma) may seek to influence behaviour and attitudes are questions that form a crucial part of the bioethical project. Bioethicists can and should ask questions about state power as well as that wielded by other influential actors.

But, what of the historical-contingent relation between power-knowledge and subjectivity that lies at the heart of the critical biopolitics? It is perhaps here that bioethics and biopolitics really do diverge because although we can see how the insights of biopolitics conceived as bioanalytics are relevant to the bioethical project, maybe bioethicists need to appreciate this point more fully. The following propositions capture an important aspect of why this may be so:

1. Subjectivity: the way people understand themselves changes over time.
2. Self-understanding affects what people consider, normal, natural, and ethical.
3. Self-understanding of individuals and populations is a source of the norms and values that bioethicists deem an important part of their methodological approach.
4. In order to affect positive change, it is important to know this and to understand how it happens.

9.11 Conclusion

To reiterate Lemke's charge against bioethics, he argues that:

> Bioethics has narrowed the terms of public debate on the relations between life and politics since the discussion is mainly conducted in ethical terms as an argument about values. (123)

The main burden of this chapter has been to consider this claim, especially its implied criticism of bioethics. Whilst acknowledging that much bioethical activity is about values, we have seen that that is certainly not all it is about, and increasingly the agenda for bioethical inquiry is moving away from the purely philosophical and ethical concerns of its early years to encompass a much broader range of methods and questions. In this sense, Lemke seems to have a rather dated and stereo-typical concept of bioethics, which does not fit the current reality.

But, we have also seen that it is not just bioethical discourse about values that is the reason for its supposed failure as a critical discipline. It also fails to account for the "epistemological and technological foundations of life processes" and "their integration into power strategies". It further neglects an examination of who possesses the resources to use the new technologies and equally ignores social constraints and institutional expectations that individuals might experience, in making their choices, all of which, "…reduces problems to alternatives that can be treated and decided…", (122–3) whereas, an analytics of biopolitics reveals complexity, historical genesis, context and generates problems and questions, opening up new horizons whilst transgressing disciplinary borders.

Still, one might quibble with some of Lemke's criticisms, as the enlarged boundaries of bioethics have increasingly seen work that touches on Lemke's biopolitical themes. For example, questions of distributive and social justice are never far from discussions of new biotechnological possibilities, which is hardly symptomatic of a neglect of who has the resources to make use of them. Many of the virtues of Lemke's 'analytics' are to be found in the expanded version of bioethics. That version is reflexive – not only does it have an expanded agenda, but it is not afraid to inspect the very premises upon which its activities and discourses rest. But, for all that, where bioethics could learn from critical biopolitics is in the appreciation that much of what we take for granted in ethical discussion is the product of social and political processes that are historically specific and contingent. Appreciating this can, as Lemke himself concludes, help us to see how things might be different.

References

Agemben, G. 1998. *Homo sacer: Sovereign power and bare life*. Stanford: Stanford University Press.

Andorno, R. 2007. Global bioethics at UNESCO: In defence of the universal declaration on bioethics and human rights. *Journal of Medical Ethics* 33 (3): 150–154.

Beauchamp, T.L., and J. Childress. 1979. *Principles of biomedical ethics*. New York: Oxford University Press. 1st ed, 1979; 5th ed, 2001.

Chattopadhyay, S. and R. de Vries. 2008. Bioethical concerns are global, bioethics is Western. *Eubios Journal of Asian and International Bioethics* 18(4): 106–109.

Foucault, M., M. Senellart, and Collège de France. 2008. *The birth of biopolitics: Lectures at the Collège de France, 1978–79*. Basingstoke/New York: Palgrave Macmillan.

Gillon, R. 2003. Ethics needs principles—Four can encompass the rest—and respect for autonomy should be "first among equals". *Journal of Medical Ethics* 29: 307–312. doi:10.1136.

Hardt, M., and A. Negri. 2000. *Empire: The new world order*. Cambridge, MA: Harvard University Press.

Häyry, M. 2003. European values in bioethics: Why, what, and how to be used. *Theoretical Medicine and Bioethics* 24 (3): 199.

Hedgecoe, A.M. 2004. Critical bioethics: Beyond the social science critique of applied ethics. *Bioethics* 18 (2): 120–143.

Hellsten, S.K. 2008. Global bioethics: utopia or reality? Devloping World Bioethics 2008(2): 70–81.

Jonsen, A.R. 1998. *The birth of bioethics*. Oxford: OUP.

Kam-por, Y. 2002. *The alleged Asian values and their implications for bioethics fourth Asian bioethics conference: Asian bioethics in the 21st century Seoul EUBIOS*.

Lemke, T. 2011. *Biopolitics: An advanced introduction*. New York: NYU Press.

Levitt, M. 2003. Better together? Sociological and philosophical perspectives on Bioethics. In *Scratching the surface of bioethics*, ed. M. Häyry and T. Takala. Amsterdam: Rodopi.

———. 2004. Complementarity rather than integration. *Medicine, Health Care and Philosophy* 7 (1): 81.

Potter, V.R. 1971. *Bioethics: Bridge to the future*. Englewood Cliffs: Prentice-Hall.

Rendtorff, J.D., and P. Kemp. 2000. *Basic ethical principles in European bioethics and biolaw*, vol. 2, Partners' Research.

Rose, N. 2007. *The politics of life itself: Biomedicince, power and subjectivity in the twenty-first century*. Princeton: Princeton University Press.

UNESCO. 2006. *Universal declaration on bioethics and human rights*, UNESCO.

Wulff, H.R. 1994. Against the four principles: A nordic view. *Principles of Health Care Ethics*: 277–286.

MIX
Papier aus verantwortungsvollen Quellen
Paper from responsible sources
FSC® C105338

FSC
www.fsc.org

If you have any concerns about our products,
you can contact us on
ProductSafety@springernature.com

In case Publisher is established outside the EU,
the EU authorized representative is:
Springer Nature Customer Service Center GmbH
Europaplatz 3, 69115 Heidelberg, Germany

Printed by Libri Plureos GmbH
in Hamburg, Germany